Cryptosporidium Serology in Human Populations

The mission of the Awwa Research Foundation (AwwaRF) is to advance the science of water to improve the quality of life. Funded primarily through annual subscription payments from over 1,000 utilities, consulting firms, and manufacturers in North America and abroad, AwwaRF sponsors research on all aspects of drinking water, including supply and resources, treatment, monitoring and analysis, distribution, management, and health effects.

From its headquarters in Denver, Colorado, the AwwaRF staff directs and supports the efforts of over 500 volunteers, who are the heart of the research program. These volunteers, serving on various boards and committees, use their expertise to select and monitor research studies to benefit the entire drinking water community.

Research findings are disseminated through a number of technology transfer activities, including research reports, conferences, videotape summaries, and periodicals.

Cryptosporidium Serology in Human Populations

Prepared by:
Dr. Judith L. Isaac-Renton and **Dr. C.S.L. Ong**
Department of Pathology and Laboratory Medicine,
University of British Columbia
655 West 12th Avenue, Vancouver, B.C. V5Z 4R4
Dr. W.R. Bowie
Department of Infectious Diseases,
University of British Columbia
655 West 12th Avenue, Vancouver, B.C. V5Z 4R4
and
Dr. P.J. Lammie and **Dr. J.W. Priest**
Division of Parasitic Diseases
Centers for Disease Control and Prevention
Atlanta, GA 30333

Sponsored by:
Awwa Research Foundation
6666 West Quincy Avenue
Denver, CO 80235-3098

Published by the

Awwa Research Foundation
Advancing the Science of Water®

IWA Publishing

DISCLAIMER

This study was funded by the Awwa Research Foundation (AwwaRF). AwwaRF assumes no responsibility for the content of the research study reported in this publication or for the opinions or statements of fact expressed in the report. The mention of trade names for commercial products does not represent or imply the approval or endorsement of AwwaRF. This report is presented solely for informational purposes.

Published by IWA Publishing, Alliance House, 12 Caxton Street, London SW1H 0QS, UK

Telephone: +44 (0) 20 7654 5500; Fax: +44 (0) 20 7654 5555; Email: publications@iwap.co.uk;
Web: **www.iwapublishing.com**

AwwaRF report number 90915F. Originally published by AwwaRF for its subscribers in 2003.
IWA Publishing version published 2004.

ISBN 1 84339 842 7

CONTENTS

TABLES

FIGURES

FOREWORD

The Awwa Research Foundation is a nonprofit corporation that is dedicated to the implementation of a research effort to help utilities respond to regulatory requirements and traditional high-priority concerns of the industry. The research agenda is developed through a process of consultation with subscribers and drinking water professionals. Under the umbrella of a Strategic Research Plan, the Research Advisory Council prioritizes the suggested projects based upon current and future needs, applicability, and past work; the recommendations are forwarded to the Board of Trustees for final selection. The foundation also sponsors research projects through the unsolicited proposal process; the Collaborative Research, Research Applications, and Tailored Collaboration programs; and various joint research efforts with organizations such as the U.S. Environmental Protection Agency, the U.S. Bureau of Reclamation, and the Association of California Water Agencies.

This publication is a result of one of these sponsored studies, and it is hoped that its findings will be applied in communities throughout the world. The following report serves not only as a means of communicating the results of the water industry's centralized research program but also as a tool to enlist the further support of the nonmember utilities and individuals.

Projects are managed closely from their inception to the final report by the foundation's staff and a large cadre of volunteers who willingly contribute their time and expertise. The foundation serves a planning and management function and awards contracts to other institutions such as water utilities, universities, and engineering firms. The funding for this research effort comes primarily from the Subscription Program, through which water utilities subscribe to the research program and make an annual payment proportionate to the volume of water they deliver and consultants and manufacturers subscribe based on their annual billings. The program offers a cost-effective and fair method for finding research in the public interest.

A broad spectrum of water supply issues is addressed by the foundation's research agenda: resources, treatment and operations, distribution and storage, water quality and analysis, toxicology, economics, and management. The ultimate purpose of the coordinated effort is to assist water suppliers to provide the highest possible quality of water economically and reliably.

The true benefits are realized when the results are implemented at the utility level. The foundation's trustees are pleased to offer this publication as a contribution toward that end.

At this time, there is little known about the contribution of drinking water to the endemic level of cryptosporidiosis. Since the USEPA is committed to proposed regulations that pertain to *Cryptosporidium*, the need for reliable epidemiological information on *Cryptosporidium* infection is critical. Epidemiological studies have been performed that utilize serologic assays as a tool for measuring infection, but the results of these studies are difficult to analyze since the nature of the antibody response to *Cryptosporidium* has not been fully characterized. This report presents results of a project designed to focus on testing and validating *Cryptosporidium*-specific serologic assays by evaluating their sensitivity and some aspects of specificity with samples collected from human populations in outbreak communities. The study also determined the duration of IgG and IgA antibody responses after exposure to *Cryptosporidium* and the prevalence of IgG *Cryptosporidium* antibody in outbreak and non-outbreak communities.

Edmund G. Archuleta, P.E.
Chair, Board of Trustees
Awwa Research Foundation

James F. Manwaring, P.E.
Executive Director
Awwa Research Foundation

ACKNOWLEDGMENTS

The authors of this report are indebted to the following water utilities and individuals for their cooperation and participation in this project:

Greater Vancouver Regional District, Quality Control, Kingsway, Burnaby, British Columbia,Canada, Robert J. Jones and Judy Smith

Capital Regional District Water Department, Water Quality Division, Island Highway, Victoria, British Columbia, Canada, Stewart Irwin

In addition, we acknowledge the valuable contribution of AwwaRF project manager, Misha Hasan, and of the members of the Project Advisory Committee (PAC), which included:

Arthur Ashendorff, New York City Department of Environmental Protection

Roger Edwards, City of Portland, Ore., Bureau of Water Works

David Matson, Eastern Virginia Medical School

The authors wish to acknowledge the technical assistance of the following individuals:

Anna Li, Pathology and Laboratory Medicine, University of British Columbia, and BCCDC Laboratory Services, Vancouver, British Columbia, Canada

James Kwon, Immunology Branch, Division of Parasitic Diseases, Centers for Disease Control, Atlanta, Georgia, USA

Mohamad K. Khan, Pathology and Laboratory Medicine, University of British Columbia

Chuck Paltiel, Statistical Consulting and Research Laboratory, Department of Statistics, University of British Columbia

Terry Peace, Public Health Protection Branch, Ministry of Health and Ministry Responsible for Seniors, Province of British Columbia

EXECUTIVE SUMMARY

Background

Cryptosporidiosis is a common infection world-wide (Hayes *et al.* 1989, Richardson *et al.* 1991, Leland *et al.* 1993, Isaac-Renton *et al.* 1994, MacKenzie *et al.* 1994). Both host and parasite factors contribute to this high prevalence. Parasite factors known to be important in transmission include: immediately infective oocysts passed in large numbers in the feces; oocysts survive in the environment for long periods of time; some species and some genotypes within species infect different kinds of hosts: and the inoculum needed to infect new hosts is low though probably variable, strain to strain. Our understanding of issues such as correlation between genotype and virulence is incomplete.

It has also been observed that contamination of raw surface and some treated drinking water supplies is widespread (Hayes *et al.* 1989, LeChevallier *et al.* 1991, Leland *et al.* 1993). The challenge to persons managing drinking water supplies is that, on one hand, contamination of public water supplies has resulted in large waterborne epidemics. This includes the Milwaukee, Wisconsin outbreak (MacKenzie *et al.* 1994) where over 400,000 persons were infected. On the other hand, the number of recognized waterborne outbreaks is less than might be expected in light of the widespread detection of *Cryptosporidium* oocysts in drinking water. It has been suggested that because prevalence is high, population immunity plays a role in limiting epidemic spread.

Host immunity probably has a role in limiting the number of epidemics as well as symptoms of disease in an individual. Moss *et al* (1998) proposed that antibody to the 15-17 and 27 kDa parasite antigens is protective against symptoms of cryptosporidiosis. Infection by *Cryptosporidium* is also more common in children (Gregory 1990), an observation consistent with the acquisition of post-exposure immunity (Current *et al.* 1983, Tzipori 1988). Thus immunity at a population level (herd immunity) may be an important factor in determining the dynamics of endemic, as well as epidemic, spread.

Immunology of *Cryptosporidium* Infection

Immunologic mechanisms responsible for either acquired or innate resistance to *Cryptosporidium* infection are not yet clearly understood. For many animal species, susceptibility to *C. parvum* infection is age-dependent. In mice, for example, neonatal animals are susceptible to infection, but by about three weeks of age, resistance to infection has developed. Humans are somewhat unusual in that naive adults are susceptible to infection. Development of resistance to repeated *Cryptosporidium parvum* infection in humans is assumed to occur, although there is only limited and indirect evidence for this. In the developing world, for example, under conditions where repeated exposure to *Cryptosporidium* is the norm, symptomatic infections are common in children, but relatively rare in adults. Further, in studies of humans who participated in volunteer studies, antibody responses directed against 27 and 17 kDa *Cryptosporidium* antigens were studied and results were found to be consistent with some degree of protective immunity following infection; however, whether antibody responses are directly involved in mediating protective immunity or serve as a marker for some other protective immune response is uncertain.

Evidence, to date, implicates a potentially protective role for both humoral and cellular immunity in *Cryptosporidium* infections. Colostrum, a potentially rich source of anti-*Cryptosporidium* antibodies, has been shown to mediate protective effects in both experimental (Fayer *et al.* 1989, 1990, Doyle *et al.* 1993)] and human infections (Tzipori *et al.* 1986, Ungar *et al.* 1990a, Nord *et al.* 1990). Studies in a murine model of cryptosporidiosis, however, suggest that antibody responses are not necessary to clear infections. *Cryptosporidium* infection in mice rendered B-cell deficient by treatment with anti-μ antibody does not differ in severity or duration from infection in control mice with a normal complement of B cells (Taghi-Kilani *et al.* 1990). That T cells play a role in controlling human infection with *Cryptosporidium* can be inferred from epidemiologic evidence demonstrating increased susceptibility of persons with decreased CD4 counts (Flanigan *et al.* 1992). Experimental evidence establishing a role for CD4 cells in the control of infection also has been obtained in murine models (Ungar *et al.* 1990b, Ungar *et al.* 1991, Mead *et al.* 1991, Zu *et al.* 1992). When administered to athymic or SCID mice, anti-interferon-γ antibody increased infection load and susceptibility to infection directly, implicating non-T cells as a source for this cytokine (Ungar *et al.* 1991, Chen *et al.* 1993). Interferon-γ knock out mice also are more susceptible to infection (Griffiths *et al.* 1998, Smith *et al.* 2000). Thus, multiple effector mechanisms are likely to play complementary roles in the control of parasite replication.

Serological Methods

Fecal examination is the routine laboratory method for diagnosis of cryptosporidiosis. Fecal examination is, however, not a practical epidemiological tool. Preliminary studies suggest that a serological approach may be useful. Anti-*Cryptosporidium* antibodies have been detected using several techniques (Ungar *et al.* 1986, Moss *et al.* 1994, Priest *et al.* 1999) and some epidemiological studies have been carried out using serological methods (Ungar *et al.* 1989, Zu *et al.* 1994, Petry 1998, Frost *et al.* 1998a, Priest *et al.* 1999 and Isaac-Renton *et al.* 1999). In one of the first studies looking at the influence of anti-*Cryptosporidium* antibodies on the outcome of infection in humans, persons infected with *Cryptosporidium* had antibody responses as demonstrated by Western blot. Using specimens from cases associated with the Milwaukee waterborne outbreak (Moss *et al.* 1994), serum antibody kinetics and isotype specificity were examined by enzyme-linked immunoelectrotransfer blot (EITB). Immunoblot formats, however, are expensive, labour-intensive, technically difficult to do and are not suitable for high throughput. The enzyme linked immunosorbent assay (ELISA) format does not have these limitations. A first generation ELISA was used to detect circulating antibodies to *Cryptosporidium* in immunocompetent and immunocompromised persons (Ungar *et al.*1986). Recently a second generation ELISA (Priest *et al.* 1999) was used to detect immunoglobulin G (IgG) to two *Cryptosporidium* sporozoite surface antigens (apparent molecular mass 17 and 27 kDa).

Relevance of Measuring Antibody Responses to the 17 and 27 kDa Antigens

As noted, the immunoblot technique has been used to characterize the antibody response of persons exposed to *Cryptosporidium* in outbreak settings to antigens derived from the oocyst stage of the parasite. A consistent finding of these studies is that two antigens of 17 and 27 kDa represent major targets of the IgG and IgA antibody response mounted by infected persons. IgA and IgG responses to both the 17 and 27 kDa antigens develop within 2 to 3 weeks after infection and persist for weeks and months, respectively. Monoclonal antibodies directed against the 17 and 27 kDa antigens localize these proteins on the surface of sporozoites, the infective stage of the parasite. The location of the antigens on the surface of the parasite is consistent with the possibility that these antigens play an important role in the process by which sporozoites attach to and invade intestinal epithelial cells. Experimental evidence supporting such a role has been obtained from studies of animals and humans. In animal models of infection, hyperimmune bovine colostrum containing high levels of sporozoite-reactive antibodies protects animals from infection. Orally administered monoclonal antibodies against the 17 and 27 kDa antigens block infection as well. In studies of humans who participated in volunteer studies, antibody responses directed against the 17 and 27 kDa antigens were associated with protection from diarrhea and with lower levels of oocyst excretion; thus, some degree of protective immunity is elicited by infection in humans. Overall, these results suggest that the 17 and 27 kDa antigens are important targets of the host immune response.

Waterborne Outbreaks of Cryptosporidiosis in British Columbia

Over 1,000 cases of cryptosporidiosis were reported to public health authorities in the western Canadian province of British Columbia during 1996. This unprecedented number of cases was associated with waterborne outbreaks occurring in this province. Health units in the south-central and south-eastern parts of the province reported the highest number of cases, with four cities in these regions - Cranbrook, Kelowna, Kamloops and Penticton - experiencing large outbreaks. Public health authorities carried out investigations of each outbreak. The largest, in

Kelowna, was associated with the use of municipal water originating from Okanagan Lake. In both Cranbrook and Penticton, case-control studies found associations between cryptosporidiosis and surface water supplies in municipal water systems. Municipal water did not appear to play a role in the fourth outbreak in Kamloops. In this outbreak, cases were more likely to have traveled to Kelowna and to have swum in a public swimming pool than did controls. As expected, children were most commonly infected. It was recognized that collection of samples for study of the host immune response would be important. Sera were collected from laboratory-confirmed cases associated with these outbreaks.

Rationale

Seroprevalence studies have been carried out using a variety of laboratory methods (Ungar *et al.* 1989, Zu *et al.* 1994, Petry 1998, Frost *et al.* 1998a and Isaac-Renton *et al.* 1999). The Centres for Disease Control and Prevention (CDC) developed an immunoblot method to detect specific antibodies to two *Cryptosporidium* antigens. This method, however, is laborious, requires highly trained staff and cannot be used in high throughput studies. The present project proposed to validate a second-generation ELISA recently developed and described by CDC.

Project Objectives

The American Water Works Association Research Foundation (AWWARF) funded this project to validate a serological method for the detection of *Cryptosporidium* infection. The main objectives of the study were as follows:

1. To determine the sensitivity of second generation enzyme linked immunosorbent assays (ELISA) developed at the Centers for Disease Control and Prevention (CDC) and of a previously described immunoblot assay.
2. To attempt to address some issues related to the specificity of the ELISA.
3. To analyze changes in levels of antibody reactivity over time in persons with laboratory-confirmed cases of cryptosporidiosis.

4. To further assess the usefulness of the ELISA at a population level by testing sera of residents of six communities, some of which had waterborne outbreaks of cryptosporidiosis.
5. To evaluate the usefulness of saliva specimens for the detection of antibodies to *Cryptosporidium* using the ELISA assay.

Approach

Sera were first tested in a blinded fashion to show that inter-laboratory correlation was acceptable and that the ELISA technology had been successfully transferred. The sensitivity of the IgG ELISA and immunoblot assay for both antigens was calculated using test results from serum samples collected from 67 laboratory-confirmed cases of cryptosporidiosis. Sensitivity of the ELISA for detection of IgA antibodies in serum and for detection of IgG antibodies in saliva was also assessed by testing specimens from laboratory-confirmed cases.

Because oocysts of *Cryptosporidium* are ubiquitous in surface water, it was not possible to obtain serum specimens from a population known to truly be unexposed. The approach taken in this study was to assess assay cross reactivity by comparing antibody levels in sera from patients confirmed to be infected with related protozoan parasites, *Toxoplasma gondii* and *Giardia lamblia.*

Sera from 41 laboratory-confirmed cases of cryptosporidiosis who provided 2 or more sera over the 100 week study associated with a waterborne outbreak were collected sequentially and tested to assess the kinetics of the host immune response. In the final phase of work, population studies were carried out to provide preliminary data on the application of these ELISA assays. Sera were collected from residents of six communities over a 24 month period. Waterborne outbreaks which occurred in two of six communities during the study period provided an excellent opportunity to obtain additional information.

Results

As previously described (Priest *et al.* 2001), arbitrary cut-offs (above 35 and 86 units for the 17 and 27 kDa antigens respectively) were used. To calculate sensitivity of the ELISA and immunoblot IgG assays, the first serum sample collected from each of 41 Kinetics Cases and the serum samples from the other 26 outbreak-associated laboratory-confirmed cases were tested. Results for the 17 and 27 kDa *Cryptosporidium* antigens were consistent with acceptable sensitivity for antibody to 17 kDa as detected by both ELISA (82%) and immunoblot (84%). Comparison of the results showed good agreement (Kappa value 0.74, standard error 0.11). The sensitivity of the ELISA for antibody to 27 kDa was greater (94%) than that found for the immunoblot (66%). Comparison of these results showed significantly less agreement (Kappa value 0.22, standard error 0.09).

The sensitivities for both IgA assay formats were unacceptably low (21 – 57%) and this isotype was not evaluated further.

A measure of the strength of the linear relationship between the different parasite ELISA assays was carried out. Using test results of sera from the outbreak-associated acute toxoplasmosis cases, the correlation coefficient of ELISA IgG *Toxoplasma* against ELISA IgG 17 kDa *Cryptosporidium* antigen was 0.07 and the correlation coefficient of ELISA IgG *Toxoplasma* against ELISA IgG 27 kDa *Cryptosporidium* antigen was - 0.02. The correlation coefficient of ELISA IgG *Giardia* results against ELISA IgG 17 kDa *Cryptosporidium* antigen was 0.14 and for the 27 kDa antigen, 0.09. Scatter plot analysis was also consistent with a lack of correlation between levels of antibody to the protozoan parasites examined.

ELISA assays (17 and 27 kDa) were carried out using 206 serum samples collected from the 41 Kinetics Cases. Since sera were collected from Kinetics Cases at arbitrary periods (every 3 months), peak antibody concentrations could not be determined with certainty. It was noted that antibody levels to both the 17 and 27 kDa antigens declined over the 100 week serum collection period. Antibody levels to the 17 kDa antigen not only fell somewhat faster than that to the 27 kDa antigen, but the median antibody ELISA Response Units to the 17 kDa antigen fell below the described positive cut-off point at 51-60 weeks. In contrast, the median antibody

ELISA response Units to the 27 kDa antigen did not fall below cut-off at any time throughout the 100 week study period. Median antibody values to 27 kDa, however, dropped to approximately twice the cut-off level after 51-60 weeks and remained at this level for the remainder of the 100 week study period. Since antibodies to 27 kDa may persist above the limit of detection, we suggest that the 27 kDa (Cp23) ELISA may be better suited to the assessment of historical exposure while the 17 kDa may be a better choice for the detection of recent infections (Priest *et al.* 2001).

Saliva samples from thirteen laboratory-confirmed cases (68 samples) of cryptosporidiosis were tested at CDC. 7 saliva samples were positive for IgG 27 kDa, 2 were positive for 17 kDa. (The 2 positives for 17 kDa were also positive for 27 kDa). Six donors accounted for the 7 samples (2 positive samples from 1 donor) positive for IgG 27 kDa. None of these positives was from an initial sample. No parasites were detected in 104 of the 109 stool samples collected from 28 Kinetics Cases (stool samples were collected along with sera at approximately 3 monthly intervals).

The ELISA format was found to be more practical in the laboratory than the immunoblot. It also provides a more objective and easily automated quantitative test result. Based on the evaluations carried out, this format was chosen for use in the community studies.

A total of 4,094 sera collected monthly from residents of 6 study communities over 24 months were assayed by ELISA for IgG antibody to 17 and 27 kDa *Cryptosporidium* antigens. Community serological results showed that the variance of the means were non-homogenous (not independent of the mean); data were log-transformed and results analyzed. Trends over time varied from community to community as noted in Figure 3.11 to Figure 3.22. The differences between the Y axis maximum values among these figures should be noted.

It is natural to ask if data from the six communities arise from the same process. This hypothesis was tested using a one-way analysis of variance or ANOVA. The null hypothesis of no difference between overall community means was rejected at the 0.001 level for both ELISA 17 and 27 kDa antigen assays.

Comparisons were carried out for each month separately to examine the null hypothesis. This analysis was done for both antigens. Additionally, for each month, a set of simultaneous

95% confidence intervals were calculated examining the differences between each of the five surface water communities (Community 2,3,4,5,6) and the deep well community (Community 1). In order to adjust for making five simultaneous comparisons to the control community (Community 1), these confidence intervals were constructed using Dunnett's (Dunnett 1955, 1964) method.

These month-by-month comparisons between Community 1 (deep well community) and each of the other communities (Community 2-6) showed increased antibody levels at the population level in Community 2 (non-outbreak) as well as Community 3 (non-outbreak) and Community 4 (outbreak). Increased antibody levels temporarily followed the waterborne outbreaks in Community 4. Community 4 uses drinking water obtained from a deep lake intake and as described, a large waterborne outbreak was identified by public health during the study period. The Community 4 serological responses to both antigens (17 and 27 kDa) rose and fell, lagging after the peak and decline in the number of cases identified. The outbreak was noted in late June-early July. As seen in our results, the community serological response began to rise in June-July, peaked in September and then slowly declined over time. Significant differences between monthly mean log ELISA results of non-outbreak communities (Community 2, 3) and the control community (Community 1) were also observed during and after the outbreak period. This may relate to travel of residents between communities, particularly since the Community 4 outbreak occurred during the province's peak vacation time.

The serological trends observed during the two years in each of the study communities are complex. For example, while the known waterborne outbreaks were clearly seen serologically (by both IgG antibodies), the reasons for several other community serological "events" which were not related temporally to the documented outbreaks are not known.

Conclusions

When the Centres for Disease Control and Prevention (CDC) developed a second generation enzyme-linked immunosorbent assays (ELISA) for detection of antibodies to the 17 and 27 kDa immunodominant parasite antigens (Priest *et al.* 1999), this method was transferred to the British Columbia (UBC) laboratory and testing carried out to validate technology transfer. Comparison of results of the blinded testing of sera showed good correlation. The ELISA test format, a format which is used in most laboratories world-wide, was shown to be readily transferable from laboratory to laboratory.

Comparison of results of immunoblot and ELISA assays showed a higher sensitivity for the ELISA IgG 27 kDa assay (94%) when compared to the immunoblot 27 kDa (66%). Both immunoblot and ELISA 17 kDa assays showed similar and acceptable sensitivity (84% and 82% respectively). Both assays for circulating IgA showed unacceptably low levels of antibody against 17 and 27 kDa antigens (range 21-57%). The sensitivity of the ELISA assay for saliva specimens was also unacceptably low.

In discussions of the specificity of antibody tests for *Cryptosporidium*, it is important to recognise that this issue should be considered on two different levels. The specificity of a test such as the ELISA can be determined if we are using another test, e.g., the large format Western blot, as a gold standard. In this case, defining the performance (sensitivity and specificity) of the secondary test relative to the primary test is straightforward and easily accomplished. A more difficult task isdefining the specificity of the antibody response. For many infectious diseases, the specificity of the antibody response is defined by analysing the antibody response of unexposed persons, those who have no contact with the organism. For *Cryptosporidium*, this approach is not possible because it is ubiquitous and all human populations are probably exposed to infection. Thus, it may be impossible to identify a human population with no potential exposure to infection.

In the absence of a truly unexposed population, we most use other strategies to determine whether infections with other organisms contribute to the antibody response to *C. parvum*. Serum specimens collected longitudinally from persons with *Toxoplasma* or *Giardia*, two important potential sources of cross reactivity, show characteristic changes in antibody responses to antigens of the infecting agent but no change in antibody response to either the 17 or 27 kDa *C. parvum* antigens. This is the most clear-cut demonstration that these organisms do not contribute to the *Cryptosporidium*-specific antibody response; however, this approach may not be feasible for the entire universe of infections that may represent potential sources of cross reactivity.

When sera from cases of acute toxoplasmosis (some with low titres, some with intermediate and some with high anti-*Toxoplasma* titres) were tested for antibodies to *Cryptosporidium*, there was no correlation between levels of antibody against antigens of these two protozoans. This is consistent with the conclusion that there was no significant cross-reaction between these two parasites. A similar analysis was carried out on sera from persons acutely infected with *Giardia*. There was no positive correlation between levels of IgG antibody to *Cryptosporidium* and *Giardia*.

Host response to *Cryptosporidium* infection varied from individual to individual, although all persons acquiring infection demonstrated a seropositivity that decreased over time. ELISA IgG to the 17 kDa antigen declined to below cut-off values at approximately 51-60 weeks post-onset. IgG response to the 27 kDa declined but did not fall below the cut-off value during the 100 weeks of the study.

During 1996 a waterborne outbreak occurred in Community 4. The number of cases reported to public health started to rise in late May, peaking in July. Serum specimens drawn monthly from this population showed a rise and fall in the community levels of antibody to both 17 and 27 kDa antigens. This community serological response lagged the epidemic curve by two months.

Analysis of community ELISA data showed significant differences across all communities. When these differences were analyzed by comparing month-by-month results, effects of the outbreak were clearly seen in the outbreak community (Community 4) and also in non-outbreak communities (Community 2, 3) as well. Other serologically identifiable events, not related in time to the known outbreaks, were also seen. We propose that within each community, complex patterns of ongoing transmission are occurring. Further, we suggest that the serological events, identified in some communities, indicate unidentified transmission events occurring within these communities. Both 17 and 27 kDa community antibody levels were significantly elevated during known waterborne outbreaks. The 17 kDa antibody levels however, appeared to detect more events in the surface water communities when compared with the deep well community.

Recommendations

The IgG ELISA assays are practical, reproducible, objective and sensitive and can be readily used to test large numbers of sera required for epidemiological studies. It is the test format of choice for further community studies.

The community serological response lagged many weeks behind the identification of the point source outbreaks. Thus, it is improbable that a serological approach will be a timely predictor of public health concerns. The 17 kDa marker is probably an indicator of more recent serological activities at the community level (Priest *et al.* 2001). Nevertheless, at the present time, both 17 and 27 kDa assays should be used in other community studies to further develop our understanding of the serological approach. When another waterborne outbreak of cryptosporidiosis occurs, it would also be useful to collect sera from volunteers for a longer period of time than 100 weeks to further study the declining host response to the 27 kDa antigen.

Future Research

More community studies using these practical and sensitive ELISA assays are needed to enhance our understanding of the parameters around a serological approach to epidemiological studies. Testing of the 1998 and 1999 sera now available to the UBC laboratory from these six study communities should be carried out. It is noted that British Columbia has a surveillance system in place for cryptosporidiosis and future studies should link serological testing with enhanced epidemiological surveillance. Expanded datasets from this further work could result in mathematical models for use in risk assessment.

CHAPTER 1
INTRODUCTION AND BACKGROUND

Background

Cryptosporidiosis is a common infection world-wide (Hayes *et al.* 1989, Richardson *et al.* 1991, Leland *et al.* 1993, Isaac-Renton *et al.* 1994, MacKenzie *et al.* 1994). Both host and parasite factors contribute to this high prevalence. Parasite factors known to be important in transmission include: immediately infective oocysts are passed in large numbers in the feces; oocysts survive in the environment for long periods of time; some species and some genotypes within species infect different kinds of hosts: and the inoculum needed to infect new hosts is low though probably variable, strain to strain. Our understanding of issues such as correlation between genotype and virulence is incomplete.

It has also been observed that contamination of raw surface and some treated drinking water supplies is widespread (Hayes *et al.* 1989, LeChevallier *et al.* 1991, Leland *et al.* 1993). The challenge to persons managing drinking water supplies is that, on one hand, contamination of public water supplies has resulted in large waterborne epidemics. This includes the Milwaukee, Wisconsin outbreak (MacKenzie *et al.* 1994) where over 400,000 persons were infected. On the other hand, the number of recognised waterborne outbreaks is less than might be expected in light of the widespread detection of *Cryptosporidium* oocysts in drinking water. It has been suggested that because prevalence is high, population immunity plays a role in limiting epidemic spread.

Host immunity probably has a role in limiting the number of epidemics as well as symptoms of disease in an individual. Moss *et al* (1998) proposed that antibody to the 15-17 and 27 kDa antigens is protective against symptoms of cryptosporidiosis. Infection by *Cryptosporidium* is also more common in children (Gregory 1990), an observation consistent with the acquisition of post-exposure immunity (Current *et al.* 1983, Tzipori 1988). Thus immunity at a population level (herd immunity) may be an important factor in determining the dynamics of endemic as well as epidemic spread.

Immunology of *Cryptosporidium* Infection

Immunologic mechanisms responsible for either acquired or innate resistance to *Cryptosporidium* infection are not yet clearly understood. For many animal species, susceptibility to *C. parvum* infection is age-dependent. In mice, for example, neonatal animals are susceptible to infection, but by about three weeks of age, resistance to infection has developed. Humans are somewhat unusual in that naive adults are susceptible to infection (Dupont *et al.* 1995). Development of resistance to repeated *Cryptosporidium parvum* infection in humans is assumed to occur, although there is only limited and indirect evidence for this. In the developing world, for example, under conditions where repeated exposure to *Cryptosporidium* is the norm, symptomatic infections are common in children, but relatively rare in adults. Further, in studies of humans who participated in volunteer studies, antibody responses directed against 27 and 17 kDa *Cryptosporidium* antigens were associated with protection from diarrhea and with lower levels of oocyst excretion. These results are consistent with the idea that some degree of protective immunity is elicited by infection in humans; however, whether antibody responses are directly involved in mediating protective immunity or serve as a marker for some other protective immune response is uncertain.

Evidence, to date, implicates a potentially protective role for both humoral and cellular immunity in *Cryptosporidium* infections. Colostrum, a potentially rich source of anti-*Cryptosporidium* antibodies, has been shown to mediate protective effects in both experimental (Fayer *et al.* 1989, 1990, Doyle *et al.* 1993)] and human infections (Tzipori *et al.* 1986, Ungar *et al.* 1990a, Nord *et al.* 1990). Studies in a murine model of cryptosporidiosis, however, suggest that antibody responses are not necessary to clear infections. *Cryptosporidium* infection in mice rendered B-cell deficient by treatment with anti-μ antibody does not differ in severity or duration from infection in control mice with a normal complement of B cells (Taghi-Kilani *et al.* 1990). That T cells play a role in controlling human infection with *Cryptosporidium* can be inferred from epidemiologic evidence demonstrating increased susceptibility of persons with decreased CD4 counts (Flanigan *et al.* 1992). Experimental evidence establishing a role for CD4 cells in the control of infection also has been obtained in murine models (Ungar *et al.* 1990b, Ungar *et al.* 1991, Mead *et al.* 1991, Zu *et al.* 1992). When administered to athymic or SCID mice, anti-interferon-γ antibody increased infection load and susceptibility to infection directly, implicating non-T cells as a source for this cytokine (Ungar *et al.* 1991, Chen *et al.* 1993). Interferon-γ knock out mice also are more susceptible to infection (Griffiths *et al.* 1998, Smith *et al.* 2000). Thus, multiple effector mechanisms are likely to play complementary roles in the control of parasite replication.

Serological Methods

Fecal examination is the routine laboratory method for diagnosis of cryptosporidiosis. Fecal examination is, however, not a practical epidemiological tool. Preliminary studies suggest that a serological approach may be useful. Anti-*Cryptosporidium* antibodies have been detected using several techniques (Ungar *et al.* 1986, Moss *et al.* 1994, Priest *et al.* 1999) and some epidemiological studies have been carried out using serological methods (Ungar *et al.* 1989, Zu *et al.* 1994, Petry 1998, Frost *et al.* 1998b, Priest *et al.* 1999 and Isaac-Renton *et al.* 1999). In one of the first studies looking at the influence of anti-*Cryptosporidium* antibodies on the outcome of infection in humans, persons infected with *Cryptosporidium* had antibody responses as demonstrated by Western blot. Using specimens from cases associated with the Milwaukee waterborne outbreak (Moss *et al.* 1994), serum antibody kinetics and isotype specificity were examined by enzyme-linked immunoelectrotransfer blot (EITB). Immunoblot formats, however, are expensive, labour-intensive, technically difficult to do and are not suitable for high throughput. The enzyme linked immunosorbent assay (ELISA) format does not have these limitations. A first generation ELISA was used to detect circulating antibodies to *Cryptosporidium* in immunocompetent and immunocompromised persons (Ungar *et al.*1986). Recently a second generation ELISA (Priest *et al.* 1999) was used to detect immunoglobulin G

(IgG) to two *Cryptosporidium* sporozoite surface antigens (apparent molecular mass 17 and 27 kDa).

Relevance of Measuring Antibody Responses to the 17 and 27 kDa Antigens

As noted, the immunoblot technique has been used to characterize the antibody response of persons exposed to *Cryptosporidium* in outbreak settings to antigens derived from the oocyst stage of the parasite. A consistent finding of these studies is that two antigens of 17 and 27 kDa represent major targets of the IgG and IgA antibody response mounted by infected persons. IgA and IgG responses to both the 17 and 27 kDa antigens develop within 2 to 3 weeks after infection and persist for weeks and months, respectively. Monoclonal antibodies directed against the 17 and 27 kDa antigens localize these proteins on the surface of sporozoites, the infective stage of the parasite. The location of the antigens on the surface of the parasite is consistent with the possibility that these antigens play an important role in the process by which sporozoites attach to and invade intestinal epithelial cells. Experimental evidence supporting such a role has been obtained from studies of animals and humans. In animal models of infection, hyperimmune bovine colostrum containing high levels of sporozoite-reactive antibodies protects animals from infection. Orally administered monoclonal antibodies against the 17 and 27 kDa antigens block infection as well. In studies of humans who participated in volunteer studies, antibody responses directed against the 17 and 27 kDa antigens were associated with protection from diarrhea and with lower levels of oocyst excretion; thus, some degree of protective immunity is elicited by infection in humans (Chappell *et al.* 1999). Overall, these results suggest that the 17 and 27 kDa antigens are important targets of the host immune response.

Waterborne Outbreaks of Cryptosporidiosis in British Columbia

Over 1,000 cases of cryptosporidiosis were reported to public health authorities in the western Canadian province of British Columbia during 1996. This unprecedented number of cases was associated with waterborne outbreaks occurring in this province. Health units in the south-central and south-eastern parts of the province reported the highest number of cases, with four cities in these regions - Cranbrook, Kelowna, Kamloops and Penticton - experiencing large outbreaks. Public health authorities carried out investigations of each outbreak. The largest, in Kelowna, was associated with the use of municipal water originating from Okanagan Lake. In both Cranbrook and Penticton, case-control studies found associations between cryptosporidiosis and surface water supplies in municipal water systems. Municipal water did not appear to play a role in the fourth outbreak in Kamloops. In this outbreak, cases were more likely to have traveled to Kelowna and to have swum in a public swimming pool than did controls. As expected, children were most commonly infected. It was recognized that collection of samples for study of the host immune response would be important. Sera were collected from laboratory-confirmed cases associated with these outbreaks.

Rationale

Seroprevalence studies have been carried out using a variety of laboratory methods (Ungar *et al.* 1989, Zu *et al.* 1994, Petry 1998, Frost *et al.* 1998a and Isaac-Renton *et al.* 1999). The Centres for Disease Control and Prevention (CDC) developed an immunoblot method to detect specific antibodies to two *Cryptosporidium* antigens. This method, however, is laborious, requires highly trained staff and cannot be used in high throughput studies. The present project proposed to validate a second-generation ELISA recently developed and described by CDC.

Project Objectives

The Awwa Research Foundation (AwwaRF) funded this project to validate a serological method for the detection of *Cryptosporidium* infection. The main objectives of the study were as follows:

1. To determine the sensitivity of second generation enzyme linked immunosorbent assays (ELISA) developed at the Centers for Disease Control and Prevention (CDC) and of a previously described immunoblot assay.

2. To attempt to address some issues related to the specificity of the ELISA.
3. To analyze changes in levels of antibody reactivity over time in persons with laboratory-confirmed cases of cryptosporidiosis.
4. To further assess the usefulness of the ELISA at a population level by testing sera of residents of six communities, some of which had waterborne outbreaks of cryptosporidiosis.
5. To evaluate the usefulness of saliva specimens for the detection of antibodies to *Cryptosporidium* using the ELISA assay.

CHAPTER 2
MATERIALS AND METHODS

Ethical Review

The Clinical Screening Committee for Research Involving Human Subjects, University of British Columbia approved this project on October 30, 1996, and issued certificate C 96-0404.

Study Communities

Six communities were studied, two where outbreaks were confirmed (1,005 sera tested) and four where no outbreaks were reported (3,089 sera tested). A total of 4,094 sera from six communities were assayed using the ELISA format. Characteristics of each community included:

1. Community 1 (Com 1 - population 105,000) uses unfiltered drinking water from deep wells. No outbreaks were identified during the study period. Drinking water was tested regularly for *Cryptosporidium* oocysts and no parasites were detected.
2. Community 2 (Com 2 - population 554,000) uses unfiltered surface drinking water from a protected watershed. No outbreaks were identified during the study period and drinking water was tested regularly. Intermittent, low levels of parasites were found.
3. Community 3 (Com 3 - population 20,000) uses unfiltered surface water from an unprotected watershed. No outbreaks were identified but this community is adjacent to Community 4. Drinking water was tested regularly and intermittently high levels of parasites found.
4. Community 4 (Com 4 - population 52,000) uses unfiltered drinking water obtained through a deep lake intake. A waterborne outbreak occurred in the summer of 1996 and oocysts were identified in the drinking water at that time.
5. Community 5 (Com 5 - population 26,000) uses unfiltered surface drinking water

obtained from an unprotected watershed. A waterborne outbreak occurred in early June of 1996 and oocysts were identified in drinking water at that time.

6. Community 6 (Com 6 - population 207,000) uses unfiltered surface drinking water from a protected watershed. No outbreak of cryptosporidiosis was identified. Drinking water was tested regularly and oocysts were occasionally identified in low levels.

Specimens

Outbreak Associated Sera

232 sera were collected from 67 laboratory-confirmed, outbreak-related cases. The 67 cases included 41 Kinetics Cases (cases providing two or more sequential samples collected every three months). Kinetics Cases provided 206 sera. 26 other laboratory-confirmed, outbreak-associated cases provided a single serum sample each. Cases were infected during outbreaks in the summer of 1996. All sera were aliquoted and frozen at -20°C until testing.

Fecal and Saliva Specimens

109 fecal and 68 saliva specimens were collected from outbreak associated laboratory-confirmed cases. Saliva were collected and frozen at -20°C. Fecal samples were collected in sodium acetate-acetic acid-formalin (SAF) preservative. The 109 stool samples did not include the initial positive samples which provided the diagnosis of cryptosporidiosis.

Toxoplasmosis and Giardiasis Sera

It is widely accepted that determining the specificity of assays for antibody to *Cryptosporidium* is problematic because it is impossible to know exactly who has, and has not, been infected with *Cryptosporidium*. In an effort to approach this issue, sera from persons with laboratory-confirmed infections with *Giardia, Toxoplasma, Entamoeba, Blastocystis* or

Microsporidium were previously analyzed at the CDC by immunoblot assay for the presence of *Cryptosporidium* antibodies. In the absence of cross-reactivity between *Cryptosporidium* and any of these agents, reactivity should be no greater than background for the populations from which the samples were collected. In this CDC work, greater than 95% of samples collected from patients with *Cryptosporidium* were found to be positive by immunoblot assay. In contrast, 60% or less of sera from patients with laboratory-confirmed *Giardia, Toxoplasma, Entamoeba, Blastocystis* or *Microsporidium* infections were positive. These results lend credibility to the idea that the ELISA used in this study is specific.

To further assess assay specificity we tested sera from persons infected with related parasites for antibody to *Cryptosporidium.* Sera from 40 cases of laboratory-confirmed cases of acute toxoplasmosis (Bowie *et al.* 1997) (10 low titre, 10 intermediate titre, 10 high titre and 5 pairs of sera from acute cases) were tested. Sera from 22 acute cases of laboratory-confirmed giardiasis were also tested (Isaac-Renton, personal communication).

Community Sera

Sera collected for routine rubella monitoring from women aged 15-40 years were used to provide a cross-section of the adult population in each study community. Because these sera reflect a healthy population in the child-bearing years, bias based on age (inclusion criteria 15-40 years) and pregnancy may be present. On the other hand, virtually all Canadian women receive prenatal care including screening for immune status to rubella: thus, the population sample provided by these specimens is likely not biased by socioeconomic factors. There does not appear to be gender bias in the transmission of cryptosporidiosis or giardiasis.

Sera were collected using a sequential list generated by the diagnostic laboratory. Specimens were from females in the defined age range who were residents of one of the study communities. Representative numbers of sera were collected for January to December 1996 and for January to December 1997. A total of 4,094 community sera were collected, aliquoted, labeled and stored at -20°C until testing.

Antigens

Antigen proteins for the immunoblots were extracted from purified *Cryptosporidium parvum* (Iowa isolate) oocysts (obtained from Centers for Disease Control) by sonication,freeze-thaw cycles and centrifugation. Antigen protein concentration in the soluble fraction of the extract were measured, aliquoted and stored at -20°C. A standardized amount of protein (4 μg) was used for each gel.

The 27 kDa antigen cloned and expressed in the pGEX expression system (recombinant *Cryptosporidium* protein (Cp23)), was recovered in high yield using enzymatic cleavage and removal of the GST by affinity chromatography. This purified recombinant protein was used in the CDC's second generation ELISA. A single-step purification technique for the isolation of the 17 kDa antigen in high yield was developed at CDC. Membrane-associated forms of the 17 kDa antigen were partially purified from sonicated oocysts by Triton X-114 phase partition and were used in the ELISA assay.

The purity of the purified antigens (recombinant 27 as well as triton extract) was assessed by immunoblot. Based on silver staining and blotting profiles, the antigens were better than 99% pure.

Assays

Immunoblot (Western Blot) Procedure

Antigen proteins were extracted from purified *Cryptosporidium parvum* oocysts by sonication and centrifugation. Antigen proteins are contained in the soluble fraction of the extract. The soluble fraction proteins were removed and stored at -80°C. Aliquots were thawed for preparation of immunoblot templates.

Extracted oocyst proteins were separated by SDS-polyacrylamide gel electrophoresis (non-reducing conditions) using Mini-Protean II Electrophoresis Cell (BioRad, Richmond, CA) for 50 minutes at 200V. 15% mini-gels (4 mg protein per gel) were used to resolve the higher and lower molecular weight components. A prestained molecular weight standard (Gibco BRL, Gaithersburg, MD) was electrophoresed in a separated well.

The separated proteins on gels were electrophoretically transferred to polyvinylidene fluoride membranes (Millipore, Bedford, MA) using Transblot SD Semi-Dry Electrophoretic Transfer Cell (BioRad, Richmond, CA) for 24 minutes at 15V. Following transfer, membranes were blocked with PBS-0.3% Tween buffer for 20 minutes at room temperature.

After blocking, one membrane was reacted with 15 test sera and a pair of positive control serum pooled from positive sera in Mini-Protean Multiscreen Apparatus which has 20 evenly spaced channels (BioRad, Richmond, CA). All sera were diluted 1:50 for IgG (1:12.5 for IgA) in PBS-0.3% Tween buffer and 400 µl of diluted sera added in each channel and incubated at 4°C overnight. The membranes were then washed with PBS-0.3% Tween buffer. The bound primary antibodies were exposed to biotinylated mouse anti-human IgG monoclonal antibodies (Zymed, South San Francisco) in 1:500 dilution. The bound secondary antibodies were incubated with streptavidin alkaline phosphatase (Gibco BRL, Gaithersburg, MD) in 1:1000 dilution (1:500 dilution for IgA) for 30 minutes at room temperature. Bands were developed with 5-bromo-4-chloro-3-indolyl phosphate substrate and nitro blue tetrazolium chromagen (BioRad, Richmond, CA) for 25 minutes at room temperature. The developed immunoblot membranes were photographed. The prints of these miniblots were read visually (blinded) by four readers. For visual readings, the photographed blot was read(blinded) by four readers and presence or absence of bands in the defined area were recorded. Doubtful results were reviewed by each reader and the tests were repeated if still doubtful. Final blot results were recorded as Positive or Negative. Any band (compared to the positive control as described above) was considered positive.

ELISA IgG Serum Assay

Antigens diluted in 0.1M $NaHCO_3$ buffer (pH 9.6) were used to sensitize 96-well plates (Immulon 2 flat-bottom microtitre immunoassay plates, Dynatech Industries Inc., McLean, Va.) overnight at 4^0C. Each well contained 50 µl of either the recombinant 27-kDa antigen (0.2 µg/ml) or the Triton X-114-extracted 17 kDa antigens (0.14 to 0.28 µg/ml) (BCA protein microassay; Pierce Biotechnology Company, Rockford, Ill.). The plates were blocked with 0.3% Tween 20–PBS for 1 h at 4^0C. After a series of three washes (subsequent washes were all with 0.05% Tween 20-PBS), 50 µl aliquots of serum diluted 1:50 with wash buffer were added to all wells. All serum samples were tested in duplicate. A twofold serial dilution (1:50 to 1:12,800) of a strong positive control was used to generate a standard curve on each individual plate. Two buffer blanks, and a battery of five serum samples known by Western blot assay to be high positive for 27 or 17 kDa, low positive for 27 or 17 kDa, combined positive for 27 and 17 kDa, combined negative for 27 and 17 kDa, and a negative pool for *C. parvum* antibodies were also included on each plate. The plates were incubated for 2 hours at room temperature and then washed four times with wash buffer. A biotinylated mouse monoclonal antibody against human IgG (clone HP6017; Zymed Laboratories) (50 µl of a 1:1,000 dilution in wash buffer) was added to each well and incubated for 1 hour at room temperature. Following four washes, the wells were filled with alkaline phosphatase-labeled streptavidin (Life Technologies) (50 µl of a 1:500 dilution in wash buffer) and incubated for an additional hour at room temperature. After four washes (the final wash was for 10 min at room temperature), *p*-nitrophenylphosphate substrate was added in 3 mM $MgCl_2$ and 10% diethanolamine at pH 10 and the color was allowed to develop until the 1:50 positive control wells had reached an absorbance of about 1.5 at 405 nm. Absorbance were measured with a Molecular Devices UVmax kinetic microplate reader. Antibody levels of the unknown samples were assigned a unit value based on the nine-point positive control standard curve with a four-parameter curve fit. The 1:50 dilution of the positive control was arbitrarily assigned a value of 6,400 U. Unknown samples with absorbance values above the standard curve were diluted further and reassayed.

ELISA IgA Serum Assay

The ELISA IgA serum assay was carried out in a similar fashion as the ELISA IgG serum assay with the following modifications: Fifty microliter aliquots of serum diluted 1:10 with wash buffer were added to all wells. A twofold serial dilution (1:10 to 1:1280) of a strong positive control was used to generate a standard curve on each individual plate. Two buffer blanks, and

1:10 dilutions of three serum samples known by Western blot assay to be positive for IgA antibodies to the 27 or 17 kDa antigens and two serum samples known to be negative for *C. parvum* antibodies were also included on each plate. A biotinylated mouse monoclonal antibody against human IgA (clone GA-112; Zymed Laboratories) (50 μl of a 1:500 dilution in wash buffer) was added to each well and incubated for 1 hour at room temperature. Color was allowed to develop until the 1:10 positive control wells had reached an absorbance of about 1.5 at 405 nm. The 1:10 dilution of the positive control was arbitrarily assigned a value of 6,400 U.

ELISA IgG Saliva Assay

The saliva assay was carried out in a similar fashion as the IgG serum assay with the following modifications: Fifty microlitres of undiluted saliva were added and the plate was incubated overnight at 4^0C to maximize antibody binding. Biotinylated mouse monoclonal antibody against human IgG and alkaline phosphatase labeled streptavidin were both used at concentrations of 1:200. The standard curve was run from 1:200 to 1:5,600 which is equivalent to 1600 units to 12.5 units by the convention of the serum assay. The plate was allowed to develop until the 1:200 standard was approximately 2.5 absorbance units.

ELISA Units and Cut-Offs

Cut-off values were determined using a relative operating characteristic (ROC) (Metz 1978) (Zweig and Campbell 1993) curve to maximize sensitivity and specificity of the ELISAs as compared with the Western blot consensus results.

The ROC curve is used to visually summarize the possible set of 2X2 tables that result when a cut-off is varied from the largest to smallest possible value. Theoretically, the greater the height of the curve above the (0,0) to (1,1) "chance" line, the better the discrimination between the two groups. The ROC curve can provide the approximate range for the ideal cut-off but the exact point would be better made using the quantitative data on the line listing based on experience related to the "costs" of false negatives versus false positives. A line listing of the false positive and false negative rates versus the unit value used for a cut-off (0-300 arbitrary units) was created from the Kinetic Cases result and is provided. See appendix tables A. 1 and A. 2. For the Cp23 ELISA, cut-off values of 66 and 113 Arbitrary Units gave a true positive and true negative rate in excess of 0.79. The corresponding cut-off unit values for the 15-17 kDa ELISA were 20 to 36 Arbitrary Units. To more accurately determine a cut-off value within the range given above, large blot and ELISA assays were conducted on an unrelated set of serum samples collected from 220 individuals during a *Cryptosporidium* outbreak investigation (Priest et al., unpublished). Cut-offs of above 86 and above 35 units for the two assays, respectively, gave the optimum results for both serum sets. A positive ELISA response for antibodies to the 27 kDa antigen was thus defined in this study as one above 86 Arbitrary Units using the recombinant Cp23 (27 kDa) protein and a positive ELISA response for antibodies to the 17 kDa antigen was defined as one above 35 arbitrary units. When applied to an unrelated set of 220 serum samples, these cut-off values compared favorably with the Western blot results when the same standard curve was used (sensitivity and specificity >87% for the Triton (17 kDa) antigen assay and >89% for the Cp23 (27 kDa) antigen assay) (Priest *et al.*, unpublished).

Examination of Fecal Specimens

Fecal samples were washed twice with saline and a modified acid fast-cold Kinyoun stain used to stain the smears prepared for the microscopic examination for *Cryptosporidium* oocysts (Garcia and Bruckner 1997). A concentration of each stool sample was carried out using the formalin-ethyl acetate sedimentation procedure and each sediment was examined microscopically for nematode eggs, larvae, and protozoan cysts and oocysts.

Statistical Methods

Analyses of data were carried out using the JMP Version: 3.2.2 software; Copyright 1989-1997; SAS Institute Inc., Cary, NC, USA and S-PLUS, version 5.1 under both Unix and Linux:

S-PLUS : Copyright (c) 1988, 1999 MathSoft, Inc., S : Copyright Lucent Technologies, Inc. Version 5.1 Release 1 for Sun SPARC, SunOS 5.5 : 1999
S-PLUS : Copyright (c) 1988, 1999 MathSoft, Inc., S : Copyright Lucent Technologies, Inc. Version 5.1 Release 1 for Linux 2.0.31 : 1999

Interpretation of the Simultaneous Community Comparison Graphs

Each graph shows, for one month, a set of simultaneous 95% confidence intervals for the difference in mean log ELISA results between Community 2, 3, 4, 5, 6 and Community 1. Dunnett's adjustment was used to compensate for making the five simultaneous comparisons. The point estimates for the differences are shown as dots, while the confidence intervals are represented by horizontal lines bounded by parentheses. A vertical solid line where the difference is zero is also included in the plot in order to assist in identifying those intervals that do not include zero. If a horizontal line bounded by the parentheses does not cross the zero line, then we are 95% confident that the true difference between that location and Community 1 is different from zero.

Quality Assurance Program

Quality Assurance Program includes the critical assessment of methods, procedures, analyses, laboratory facilities and equipment as well as ongoing technical performance. The ongoing laboratory QC program includes:

1. Maintenance and daily checks with written records of equipment function. This includes daily temperature checks and logs for incubators, refrigerators and freezers; rpm checks for centrifuges; pH meter and balance checks.
2. Written detailed procedures are in place for basic laboratory methods including commercially available kits, parasitologic methods, serum retrieval procedures,

database entry procedures, and serological methods. These methods are reviewed regularly.

3. Procedures have been established to ensure that all patient information is linked to test results by assigning each sera and patient a unique coded study number. Procedures have also been setup to maintain patient confidentiality.
4. All sera are aliquoted so as to avoid freeze-thaw effect. Date of aliquoting, location in the freezer, and total volume available is recorded.
5. Detailed information about antigen source (host, supplier, number of passages, maintenance, date of origin and date of receipt in lab etc) are kept.
6. Detailed information about reagents, standards purchase, receipt and first use dates, sources and batches are recorded on an ongoing basis.

Quality control procedures specific to the serological methods included the following:

ELISA method

1. All 96 wells on the plates were used.
2. Positive and negative controls as well as a standard curve were used to control for plate to plate variation. Five pairs (10 wells) are used per plate for known controls. Sera from pooled positives are used in 6 wells per plate, sera from negative controls are in 4 wells per plate. Antibody values for the control sera, as calculated from the standard curve, were used to accept or reject the assay run. Duplicate wells with coefficient of variation greater than 15% were indications that the assay must be repeated.
3. Four wells per plate were not inoculated with serum and they to serve as reagent controls. Non-antigen containing wells were used to blank the ELISA reader for every plate.
4. Duplicate wells were run for each unknown. When well-to-well variation exceeded 15% (CV), the sample was retested.

5. Positive and negative controls were assayed to check intra- and inter-assay variation. The mean and standard deviation for both controls were calculated for at least 20 plates (different runs). These results were used to calculate exclusion criteria (anything outside the mean plus 2 SD) which excluded any plates that were outside the 95.5% frequency interval assuming a normal distribution. Also, one randomly selected unknown sera was rerun on each plate. If mean antibody values for the replicates differ by more than 10%, the entire plate was rerun. Three controls were used on each plate and included a high positive, a low positive and a combined positive control. If the test results of at least two of these controls were within the acceptable limits then all the plate results were accepted.
6. Each plate was assigned a number. Data (optical densities) were recorded on disk from the micro-reader and attached by staple to a record sheet for that plate.
7. Entry of data into a Microsoft ® Excel 2000 database was carried out and a photocopy of records were made. Technologists carrying out the work also kept a record log of each day's run.

Immunoblot method

1. Lane 1 and 20 (outside lanes) were not used. Lane 19 was used for reagent control (blank). A color-coded molecular weight standard was run on each blot.
2. Two lanes, one on each half of the blot, were used for testing positive control sera (pooled positives). Lanes were chosen randomly on either side of the blot for control sera.
3. The developed immunoblot membranes were photographed for visual reading.
4. Technologists carrying out the work kept a record log of each day's run.

CHAPTER 3
RESULTS

ELISA Inter-Laboratory Comparison

Using 232 sera (split and blind-coded) results of the ELISA assays carried out in the CDC laboratory were compared to results on the same specimens (using the same procedure and antigens) in the UBC laboratory. Figure 3.1 and Figure 3.2 are scatter plots showing correlation between the results of IgG ELISA for 17 and 27 kDa assays. Correlation coefficients between laboratories were consistent with successful technology transfer: 0.91 for both 17 (Kappa value 0.82, standard error 0.38) and 27 kDa assays (Kappa value 0.86, standard error 0.05).

ELISA Variability

Results from 98 assays for the 27 kDa antigen for both High and Low positive controls and 96 assays for the 17 kDa antigen for both High and Low positive controls were used to prepare the Shewhart Control Charts (Shewhart 1931, Westgard *et al.* 1981) as shown in Figures 3.3 and 3.4. Figure 3.5 shows results for 95 individual assays of the Combined positive control. In all tests, at least two of the three controls were within the acceptable upper and lower control limits shown or the results were rejected and the test repeated. Results were consistent with acceptable day-to-day test variability.

IgG Assay Sensitivity

To calculate sensitivity of the ELISA and immunoblot IgG assays, the first serum sample collected from each of 41 Kinetics Cases and the serum samples from the other 26 outbreak-associated laboratory-confirmed cases were tested. As previously described (Priest *et al.* 2001), arbitrary cut-offs (above 35 and 86 units for 17 and 27 kDa antigens respectively) were used.

Results for the 17 and 27 kDa *Cryptosporidium* antigens (Table 3.1) are consistent with acceptable sensitivity for antibody to 17 kDa as detected by both ELISA (82%) and immunoblot (84%). Comparison of the results showed good agreement (Kappa value 0.74, standard error 0.11). The sensitivity of the ELISA for antibody to 27 kDa was greater (94%) than that found for the immunoblot (66%). Comparison of these results showed significantly less agreement (Kappa value 0.22, standard error 0.09).

Table 3.1

Sensitivity of IgG ELISA and IgG immunoblot assay using outbreak associated cases

Cases	# Sera Tested	ELISA 27 kDa Positive (% positive)	Immunoblot 27 kDa Positive (% positive)	ELISA 17 kDa Positive (% positive)	Immunoblot 17 kDa Positive (% positive)
Laboratory confirmed	67	63 (94)	44 (66)	55 (82)	56 (84)

IgA Assay Sensitivity

To assess sensitivity of the ELISA and immunoblot assays, the first serum sample from each of 41 Kinetics Cases sera and from each of these 26 outbreak-associated, laboratory-confirmed cases were tested. Results are seen in Table 3.2. The sensitivities for both assay formats were unacceptably low (21 – 57%). This isotype was not evaluated further.

Table 3.2

Sensitivity of IgA ELISA and IgA immunoblot assay using outbreak associated cases

Cases	# Sera Tested	ELISA 27 kDa Positive (% positive)	Immunoblot 27 kDa Positive (% positive)	ELISA 17 kDa Positive (% positive)	Immunoblot 17 kDa Positive (% positive)
Laboratory confirmed	67	29 (43)	14 (21)	38 (57)	30 (45)

Assay Cross Reactivity

A measure of the strength of the linear relationship between the different parasite ELISA assays was carried out. If there is an exact linear relationship between two variables, the correlation is 1 or –1; if there is no linear relationship, the correlation tends toward zero. Using test results of sera from the outbreak-associated acute toxoplasmosis cases, the correlation coefficient of ELISA IgG *Toxoplasma* against ELISA IgG 17 kDa *Cryptosporidium* antigen was 0.07. The correlation coefficient of ELISA IgG *Toxoplasma* against ELISA IgG 27 kDa *Cryptosporidium* antigen was - 0.02. In 5 paired sera which showed a rise in antibody levels to *Toxoplasma* antigens, there were no concomitant rise in antibody levels to the *Cryptosporidium* antigens. The correlation coefficient of ELISA IgG *Giardia* results against ELISA IgG 17 kDa *Cryptosporidium* antigen was 0.14 and for the 27 kDa antigen, 0.09. Scatter plot analysis (Figures 3.6 to 3.9) was also consistent with a lack of correlation between levels of antibody to the protozoan parasites examined.

Host Immune Response Measured by ELISA Assays

ELISA assays (17 and 27 kDa) were carried out using 206 serum samples collected from the 41 Kinetics Cases. The median response (Units) of duplicate testing at each time interval post-onset are summarized graphically in Figure 3.10. Since sera were collected from Kinetics Cases at arbitrary periods (every 3 months), peak antibody concentrations could not be determined with certainty. It was noted that antibody levels to both the 17 and 27 kDa antigens declined over the 100 week serum collection period.

The ELISA cut-offs (positive above 35 Units for 17 kDa, positive above 86 Units for 27 kDa) are not shown in this Figure because of possible confusion using dual Y axes. It is clear, however, that antibody levels to the 17 kDa antigen fell somewhat faster than that to the 27 kDa antigen, and the median antibody ELISA Response Units to the 17 kDa antigen fell below the positive cut-off point at 51-60 weeks. In contrast, the median antibody ELISA response Units to the 27 kDa antigen did not fall below cut-off at any time throughout the 100 week study period. Median antibody values to 27 kDa, however, dropped to approximately twice the cut-off level after 51-60 weeks and remained at this level for the remainder of the 100 week study period.

Fecal and Saliva Test Results

Saliva samples from thirteen laboratory-confirmed cases (68 samples) of cryptosporidiosis were tested at CDC. Applying the serum cut-off values of above 35 Units for 17 kDa and above 86 Units for 27 kDa, 7 saliva samples were positive for IgG 27 kDa, 2 were positive for 17 kDa.. The 2 positives for 17 kDa were also positive for 27 kDa. Six donors accounted for the 7 samples (2 positive samples from 1 donor) positive for IgG 27 kDa. None of these positives was from an initial sample.

No parasites were detected in 104 of the 109 stool samples collected from 28 Kinetics Cases (stool samples were collected along with sera at approximately 3 monthly intervals). *Cryptosporidium* was detected in 1 specimen (approximately 6 months post-onset) in one patient. *Giardia* was detected in another 4 specimens, all from one case (no *Cryptosporidium* was detected in these 4 specimens).

Community Studies

In a previous community study (Isaac-Renton *et al.* 1999), sera were tested using an immunoblot format. These sera were assayed in this project by ELISA. The ELISA format was found to be more practical in the laboratory than the immunoblot when testing large number of specimens. The ELISA format also provides a more objective and easily automated quantitative test result. Based on the evaluations carried out, this format was chosen for use in the community studies.

Overall, a total of 4,094 sera collected monthly from residents of 6 study communities over 24 months were assayed by ELISA for IgG antibody to 17 and 27 kDa *Cryptosporidium* antigens. The results of these assays are shown in Table 3.3 to Table 3.8. The plots of the ELISA monthly mean and median values for both the 17 and 27 kDa antibody responses for 1996 and 1997 are shown in Figure 3.11 to Figure 3.22. Trends over time varied from community to community.

It is natural to ask if data from all six communities arise from the same process. This hypothesis was tested using a one-way analysis of variance or ANOVA. The null hypothesis of no difference between overall community means was rejected at the 0.001 level for both ELISA 17 and 27 kDa antigen assays. Comparisons were then carried out for each month separately, using one-way ANOVA, to examine the null hypothesis that the study communities have the same monthly mean log ELISA results. This analysis was done for both antigens. Additionally, for each month, a set of simultaneous 95% confidence intervals were calculated examining the differences between each of the five surface water communities (Community 2,3,4,5,6) and the deep well community (Community 1). In order to adjust for making five simultaneous comparisons to the control community (Community 1), these confidence intervals were constructed using Dunnett's method (Dunnett 1955, 1964). The resulting comparisons for the month of September 1996 for 17 and 27 kDa respectively are represented graphically in Figure 3.23 and Figure 3.24.

Month-by-month comparisons between Community 1 (deep well community) and each of the other communities (Community 2-6), showed significantly increased antibody levels after the waterborne outbreaks in Community 4. Community 4 uses drinking water obtained from a deep lake intake and as described, a large waterborne outbreak was identified by public health in the summer of 1996. Figure 3.25 shows the relationship between the Community 4 waterborne outbreak and the community's serological response. The peaking serological response is validated by the excess number of cases of cryptosporidiosis identified by the community health care system to public health. The community serological responses to both antigens (17 and 27 kDa) rose and fell, lagging after the peak and decline in the number of cases identified. The outbreak was noted in late June to early July, the community serological response rose in August, peaked in September and slowly declined over time. Significant differences between monthly mean log ELISA results of non-outbreak communities (Community 2, 3) and the control community (Community 1) were seen for the outbreak period. This may be related to travel of residents between communities, particularly since the Community 4 outbreak occurred during the province's peak vacation time. The community trends observed during the two years in each of the study communities are complex. For example, while known waterborne outbreaks were clearly seen serologically (by both IgG antibodies), the reasons for several other community serological "events," not related temporally to documented outbreaks, are not known.

Table 3.3
Community 1 (Com 1) 17 and 27 kDa ELISA IgG 25results for 1996 and 1997
(positive based on utilization of 36 and 87 units for 17 and 27 kDa antigens respectively)

Com 1	ELISA % Positive		ELISA Response (Units)			
Month	17 kDa % Pos	27 kDa % Pos	17 kDa Median	17 kDa Mean	27 kDa Median	27 kDa Mean
Jan 96	42	84	26	48	218	339
Feb 96	46	89	33	99	194	359
Mar 96	52	89	36	55	231	592
Apr 96	50	82	40	51	183	439
May 96	38	48	18	48	100	255
Jun 96	41	86	29	105	285	556
Jul 96	38	79	32	44	183	317
Aug 96	48	81	30	68	232	373
Sep 96	50	70	32	57	149	443
Oct 96	35	77	26	33	176	409
Nov 96	49	83	34	78	197	461
Dec 96	41	45	31	52	78	291
Jan 97	65	87	43	70	232	428
Feb 97	68	82	30	82	187	287
Mar 97	64	73	40	72	273	487
Apr 97	52	57	40	82	137	287
May 97	60	80	47	108	283	432
Jun 97	52	91	40	52	223	306
Jul 97	70	87	52	97	344	544
Aug 97	48	74	34	65	180	283
Sep 97	50	77	41	63	278	439
Oct 97	44	74	29	54	205	387
Nov 97	52	74	37	78	318	605
Dec 97	41	73	31	103	134	483

Table 3.4
Community 2 (Com 2) 17 and 27 kDa ELISA IgG results for 1996 and 1997
(positive based on utilization of 36 and 87 units for 17 and 27 kDa antigens respectively)

Com 2	ELISA % Positive		ELISA Response (Units)			
Month	17 kDa % Pos	27 kDa % Pos	17 kDa Median	17 kDa Mean	27 kDa Median	27 kDa Mean
Jan 96	71	90	72	160	333	552
Feb 96	65	79	52	97	223	355
Mar 96	73	88	55	116	288	428
Apr 96	62	79	50	83	205	338
May 96	58	80	44	89	213	438
Jun 96	60	80	47	99	200	388
Jul 96	53	87	40	106	242	515
Aug 96	73	85	48	107	227	407
Sep 96	67	83	48	108	230	442
Oct 96	70	89	62	140	255	480
Nov 96	74	85	62	133	288	465
Dec 96	61	85	46	104	222	419
Jan 97	78	96	112	166	339	599
Feb 97	96	96	79	140	395	760
Mar 97	70	96	48	76	212	284
Apr 97	61	78	40	96	179	317
May 97	61	83	41	73	253	414
Jun 97	70	96	46	79	269	431
Jul 97	83	78	47	85	176	278
Aug 97	74	83	54	87	198	304
Sep 97	87	87	79	115	177	315
Oct 97	87	78	58	90	270	332
Nov 97	70	91	65	92	282	509
Dec 97	71	95	101	106	486	409

Table 3.5
Community 3 (Com 3) 17 and 27 kDa ELISA IgG results for 1996 and 1997
(positive based on utilization of 36 and 87 units for 17 and 27 kDa antigens respectively)

Com 3	ELISA % Positive		ELISA Response (Units)			
Month	17 kDa % Pos	27 kDa % Pos	17 kDa Median	17 kDa Mean	27 kDa Median	27 kDa Mean
Jan 96	67	86	46	72	359	512
Feb 96	33	50	21	36	110	193
Mar 96	65	71	47	116	328	520
Apr 96	39	85	28	52	267	255
May 96	36	88	32	59	340	396
Jun 96	47	94	31	55	257	482
Jul 96	77	91	93	338	531	1478
Aug 96	69	81	88	353	357	745
Sep 96	67	81	68	1161	302	859
Oct 96	78	100	78	124	487	589
Nov 96	78	94	60	84	386	473
Dec 96	57	86	65	249	215	722
Jan 97	69	94	86	241	466	1068
Feb 97	75	100	201	181	1011	1187
Mar 97	100	100	98	98	1229	1229
Apr 97	78	100	73	213	561	983
May 97	86	100	105	78	596	652
Jun 97	50	100	92	92	249	249
Jul 97	100	100	107	107	413	413
Aug 97	60	80	38	45	151	234
Sep 97	80	100	77	69	244	365
Oct 97	100	100	167	167	508	468
Nov 97	50	100	46	49	317	291
Dec 97	50	100	70	111	305	279

Table 3.6
Community 4 (Com 4) 17 and 27 kDa ELISA IgG results for 1996 and 1997
(positive based on utilization of 36 and 87 units for 17 and 27 kDa antigens respectively)

Com 4	ELISA % Positive		ELISA Response (Units)			
Month	17 kDa % Pos	27 kDa % Pos	17 kDa Median	17 kDa Mean	27 kDa Median	27 kDa Mean
Jan 96	57	100	46	97	249	500
Feb 96	83	91	111	158	450	651
Mar 96	57	83	39	90	308	596
Apr 96	61	78	42	110	293	548
May 96	70	91	58	110	290	556
Jun 96	83	91	86	176	453	808
Jul 96	78	96	65	95	760	812
Aug 96	91	100	76	229	464	705
Sep 96	96	100	202	575	868	3103
Oct 96	91	91	160	244	1085	1731
Nov 96	90	100	102	179	512	820
Dec 96	87	96	62	128	312	460
Jan 97	91	96	126	153	422	646
Feb 97	74	96	63	161	475	700
Mar 97	91	87	87	212	317	487
Apr 97	83	96	92	97	385	511
May 97	78	91	106	126	317	613
Jun 97	65	74	51	160	232	495
Jul 97	74	96	52	99	364	513
Aug 97	52	96	44	89	298	408
Sep 97	52	83	40	63	128	242
Oct 97	74	87	75	128	208	370
Nov 97	91	100	71	101	330	554
Dec 97	74	91	105	145	217	342

Table 3.7

Community 5 (Com 5) 17 and 27 kDa ELISA IgG results for 1996 and 1997
(positive based on utilization of 36 and 87 units for 17 and 27 kDa antigens respectively)

Com 5	ELISA % Positive		ELISA Response (Units)			
Month	17 kDa % Pos	27 kDa % Pos	17 kDa Median	17 kDa Mean	27 kDa Median	27 kDa Mean
Jan 96	44	94	34	109	256	424
Feb 96	57	86	44	72	232	394
Mar 96	80	85	49	68	233	291
Apr 96	57	91	43	121	316	598
May 96	35	100	34	51	335	583
Jun 96	78	91	50	108	277	389
Jul 96	61	83	42	433	349	980
Aug 96	56	78	36	43	187	255
Sep 96	65	96	39	48	278	503
Oct 96	57	91	37	156	185	364
Nov 96	48	78	35	66	208	307
Dec 96	74	91	51	87	201	421
Jan 97	50	69	40	70	173	270
Feb 97	59	90	46	98	259	471
Mar 97	50	92	32	83	347	324
Apr 97	88	100	55	133	352	617
May 97	57	86	40	78	186	1633
Jun 97	60	93	65	63	276	399
Jul 97	38	100	24	85	251	590
Aug 97	86	86	73	106	375	422
Sep 97	65	96	48	79	223	655
Oct 97	63	94	39	72	259	357
Nov 97	56	88	41	79	216	280
Dec 97	64	100	61	84	229	404

Table 3.8
Community 6 (Com 6) 17 and 27 kDa ELISA IgG results for 1996 and 1997
(positive based on utilization of 36 and 87 units for 17 and 27 kDa antigens respectively)

Com 6	ELISA % Positive		ELISA Response (Units)			
Month	17 kDa % Pos	27 kDa % Pos	17 kDa Median	17 kDa Mean	27 kDa Median	27 kDa Mean
Jan 96	61	96	40	65	231	364
Feb 96	48	83	32	62	229	285
Mar 96	57	87	61	112	256	306
Apr 96	57	87	40	140	220	333
May 96	44	74	33	61	169	291
Jun 96	74	83	45	83	182	233
Jul 96	83	65	42	97	112	265
Aug 96	74	87	48	68	139	230
Sep 96	48	78	33	80	137	187
Oct 96	39	74	31	44	154	200
Nov 96	52	83	39	607	279	359
Dec 96	63	95	41	53	153	274
Jan 97	52	83	48	62	229	301
Feb 97	59	91	38	59	251	353
Mar 97	48	81	33	40	141	214
Apr 97	65	74	41	53	294	323
May 97	48	78	32	65	209	350
Jun 97	64	77	47	92	198	605
Jul 97	55	86	42	79	205	482
Aug 97	52	70	38	89	197	401
Sep 97	74	78	60	87	173	316
Oct 97	74	83	55	176	220	466
Nov 97	52	70	39	79	126	389
Dec 97	62	95	49	88	406	566

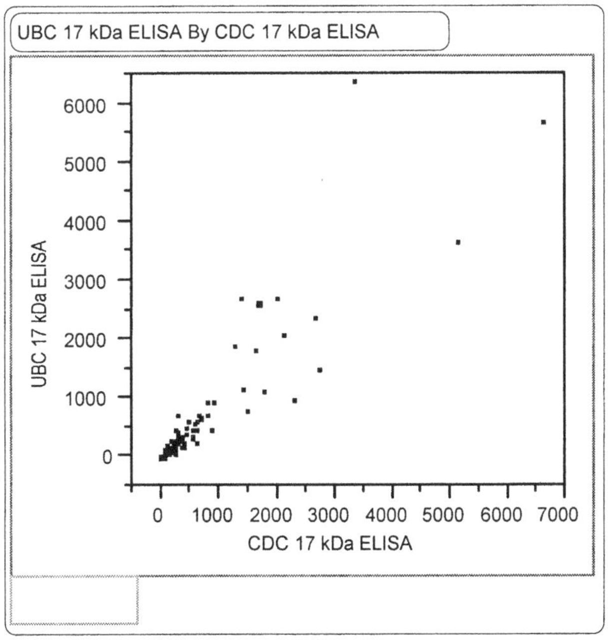

Figure 3.1 Inter laboratory comparison scatter plot of ELISA IgG 17 kDa *Cryptosporidium* assay results for the same sample set (n=232, Corr. Coef. = 0.91)

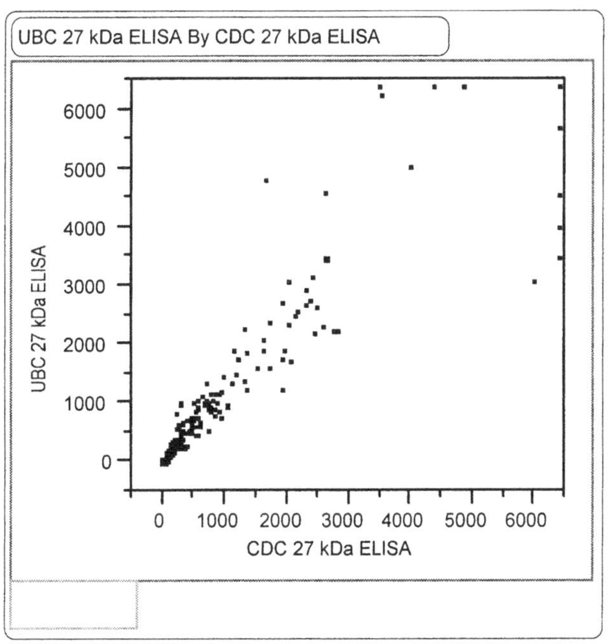

Figure 3.2 Inter laboratory comparison scatter plot of ELISA IgG 27 kDa *Cryptosporidium* assay results for the same sample set (n=232, Corr. Coef. = 0.91)

(a)

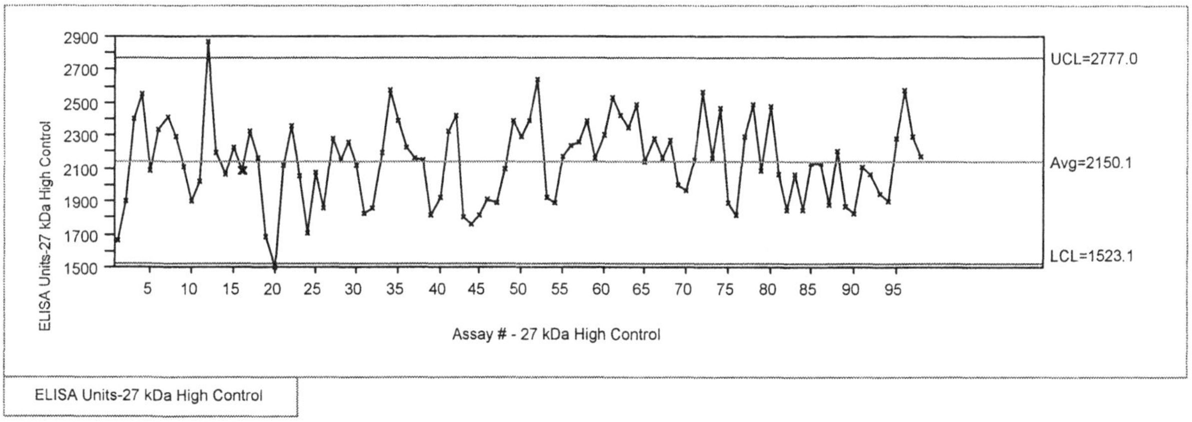

(b)

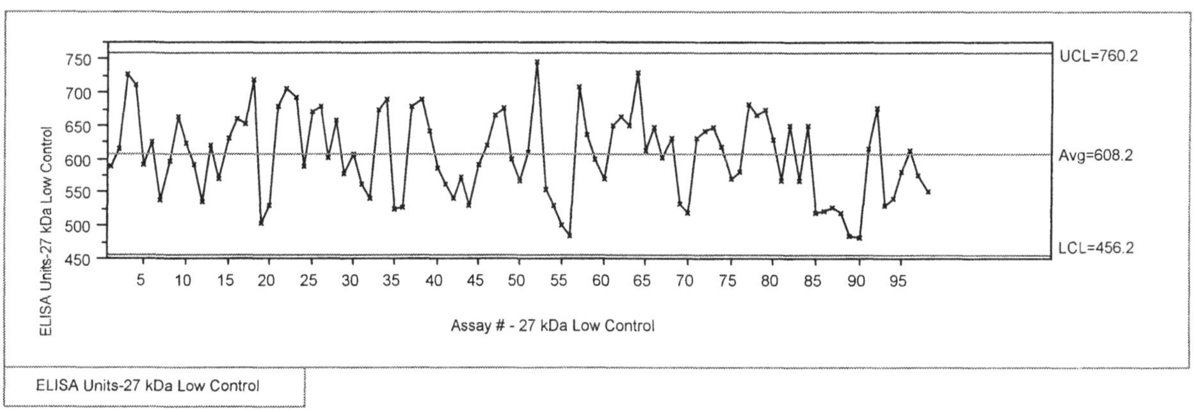

Figure 3.3 ELISA variability for 27 kDa High (a) and Low (b) positive control sera

(a)

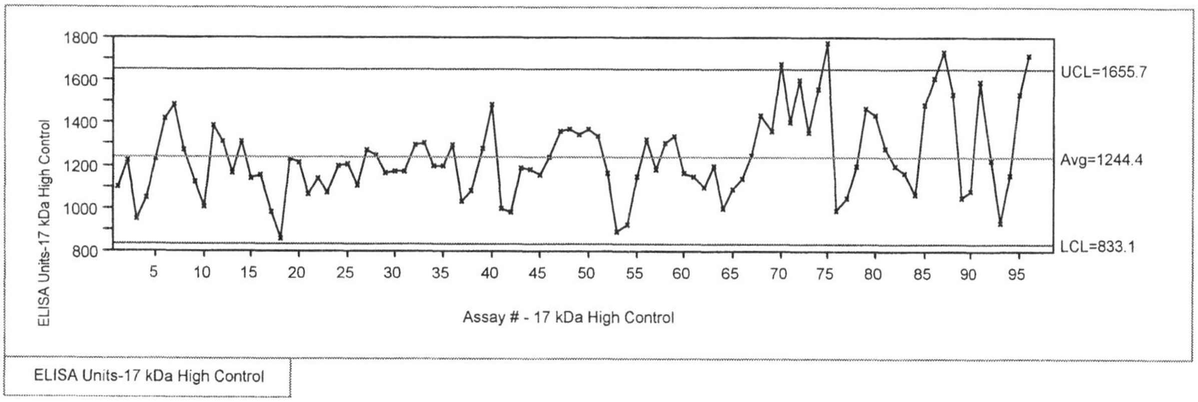

(b)

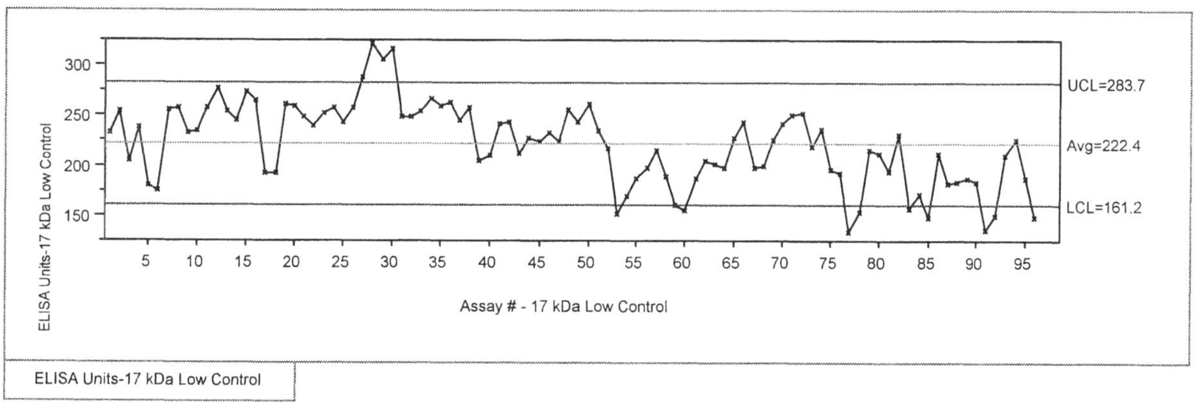

Figure 3.4 ELISA variability for 17 kDa High (a) and Low (b) positive control sera

(a)

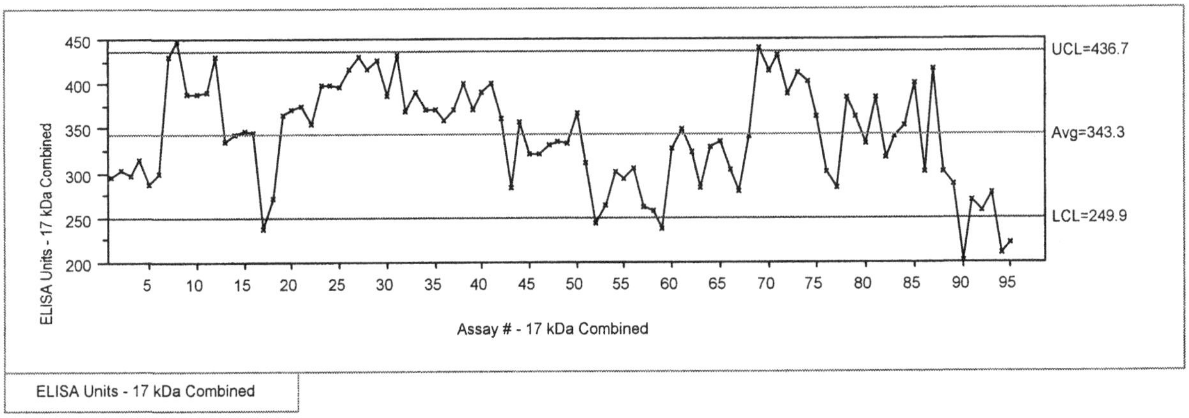

(b)

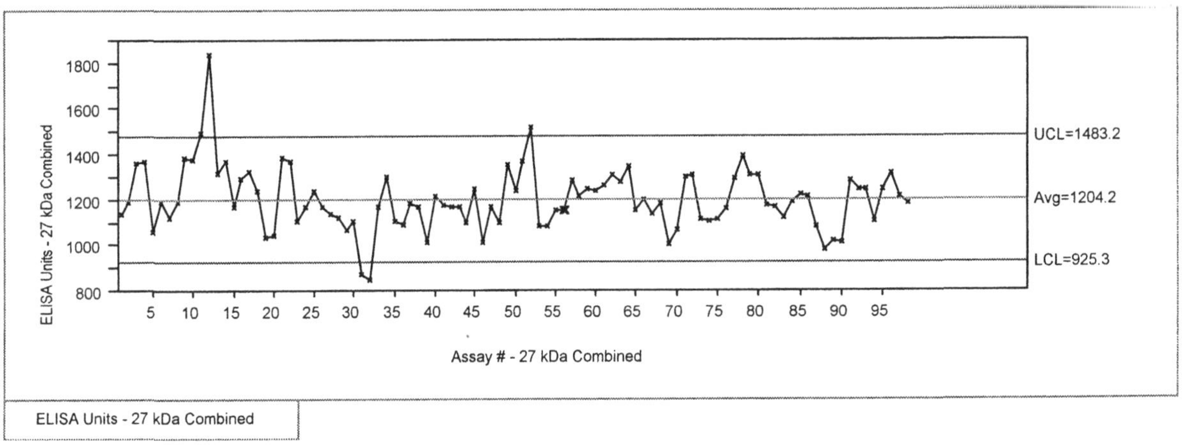

Figure 3.5 ELISA variability for 17 kDa (a) and 27 kDa (b) Combined positive control sera

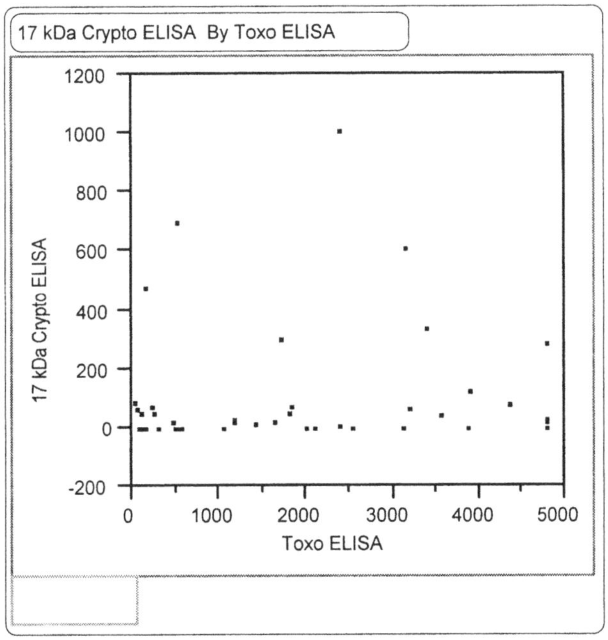

Figure 3.6 Scatter plot of ELISA IgG 17 kDa *Cryptosporidium* assay results and ELISA IgG *Toxoplasma* assay results for the same sera (Corr. Coef. = 0.07)

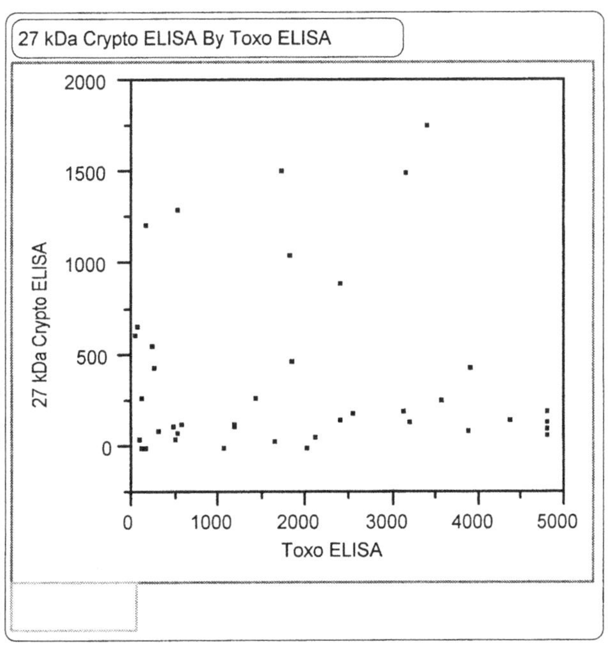

Figure 3.7 Scatter plot of ELISA IgG 27 kDa *Cryptosporidium* assay results and ELISA IgG *Toxoplasma* assay results for the same sera (Corr. Coef. = -0.02)

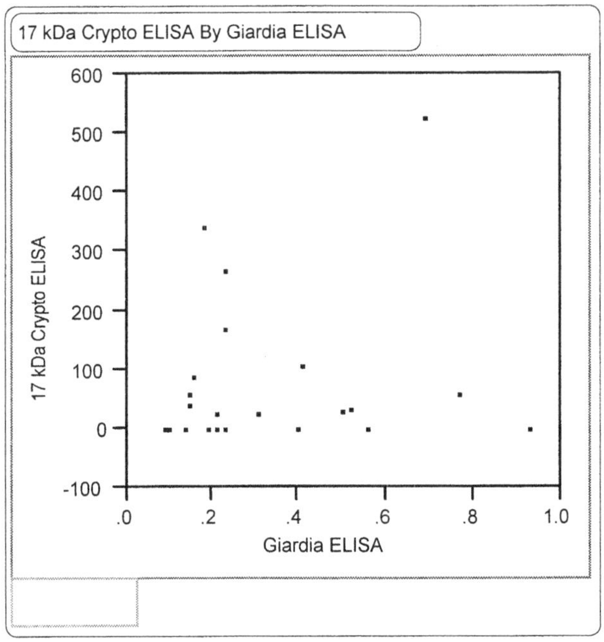

Figure 3.8 Scatter plot of ELISA IgG 17 kDa *Cryptosporidium* assay results and ELISA IgG *Giardia* assay results for the same sera (Corr. Coef. = 0.14)

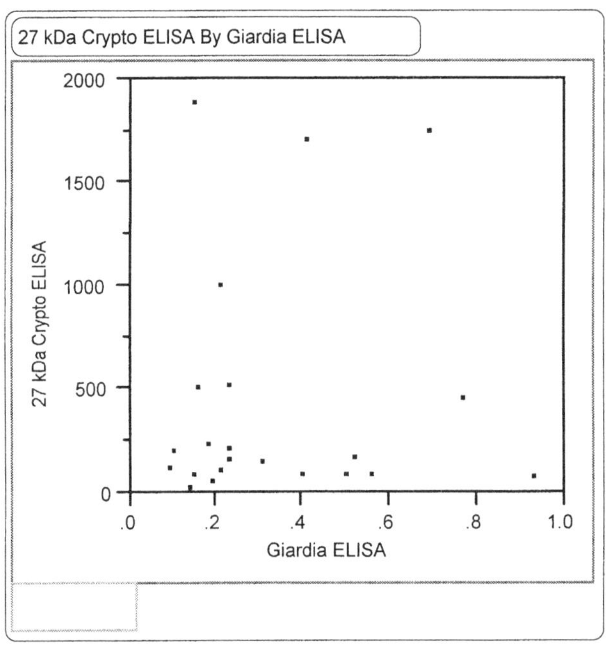

Figure 3.9 Scatter plot of ELISA IgG 27 kDa *Cryptosporidium* assay results and ELISA IgG *Giardia* assay results for the same sera (Corr. Coef. = 0.09)

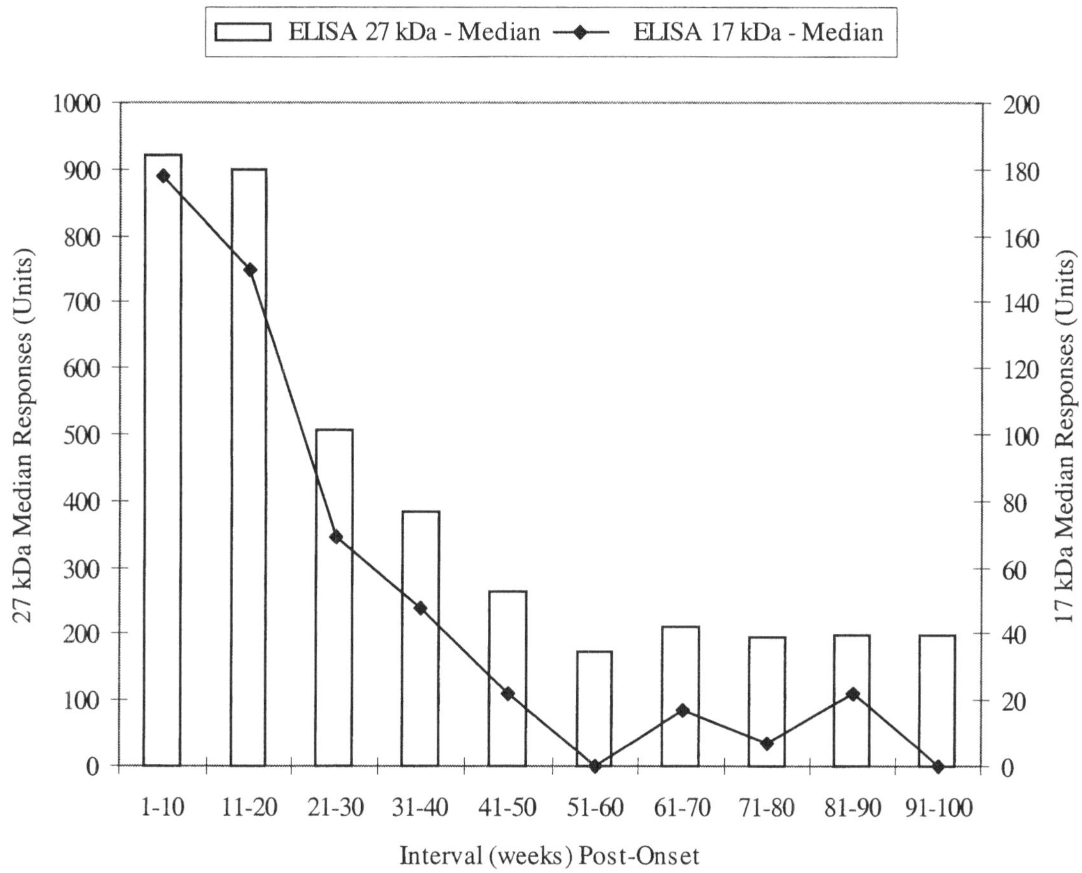

Figure 3.10 Median responses (Units) for Kinetics Cases obtained using the ELISA IgG 17 and 27 kDa assays. Sera were obtained at 3 month time intervals post-onset from laboratory-confirmed outbreak associated cases

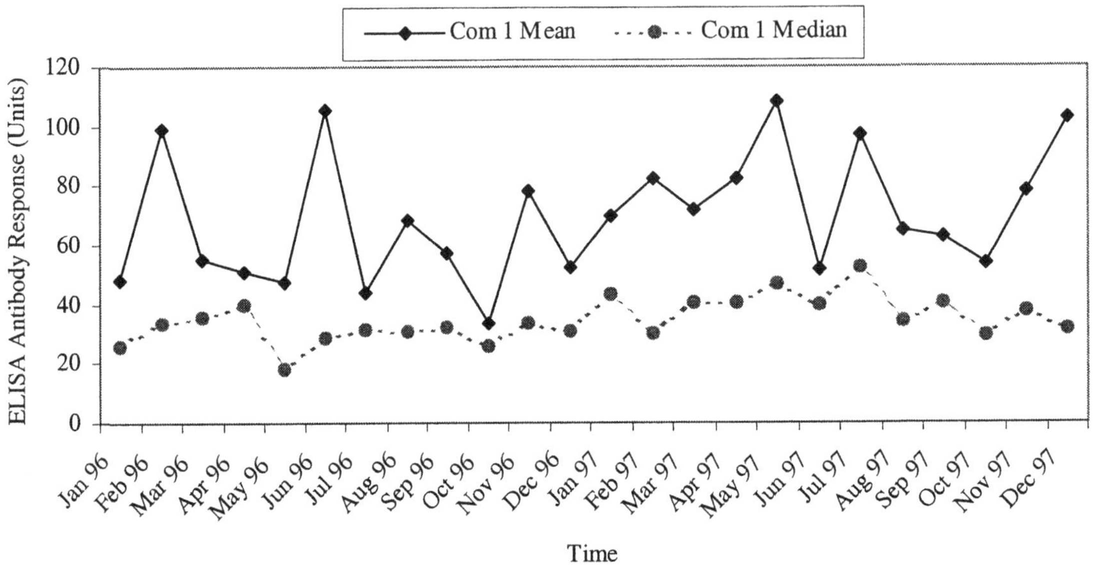

Figure 3.11 Community 1 - ELISA monthly mean and median for 17 kDa
(note differences in maximum y axis values between figures)

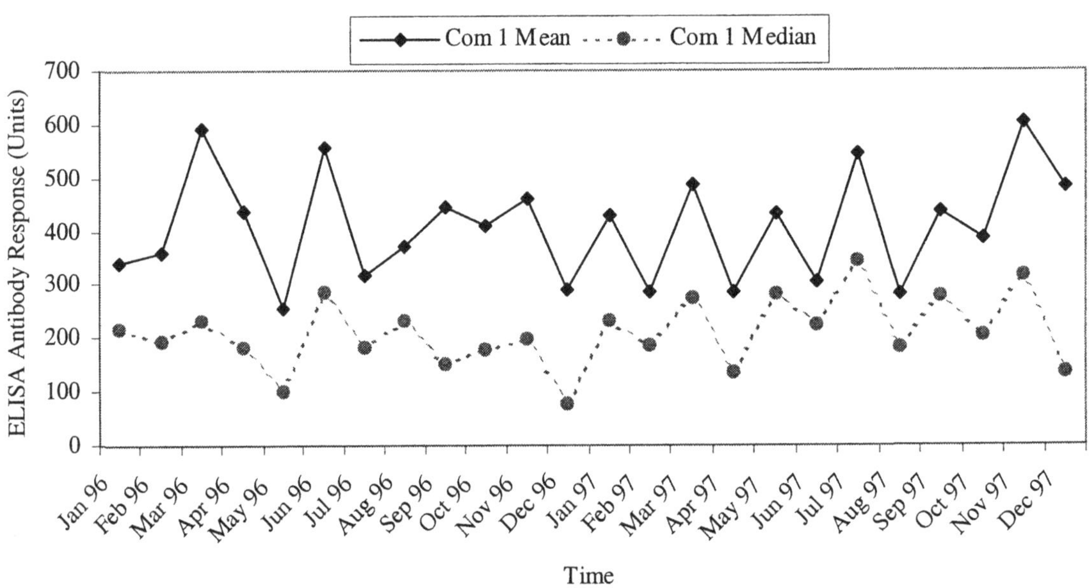

Figure 3.12 Community 1 - ELISA monthly mean and median for 27 kDa
(note differences in maximum y axis values between figures)

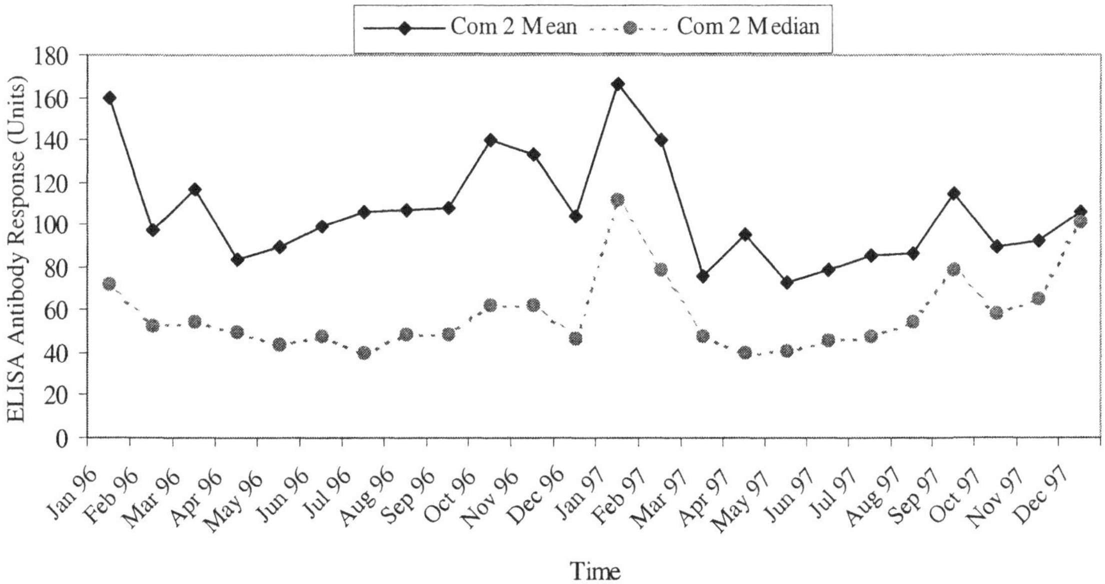

Figure 3.13 Community 2 - ELISA monthly mean and median for 17 kDa
(note differences in maximum y axis values between figures)

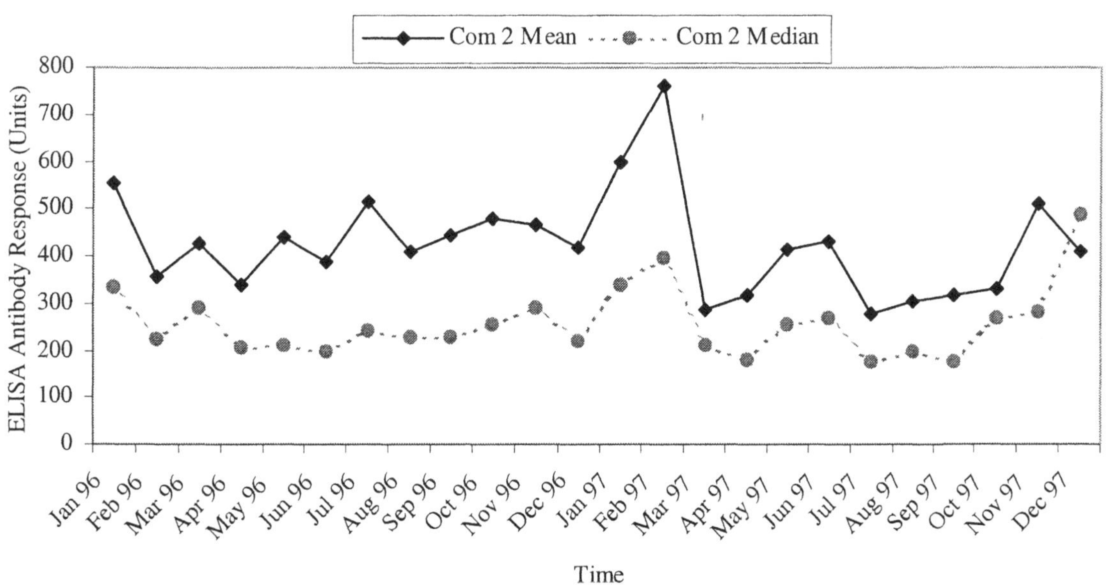

Figure 3.14 Community 2 - ELISA monthly mean and median for 27 kDa
(note differences in maximum y axis values between figures)

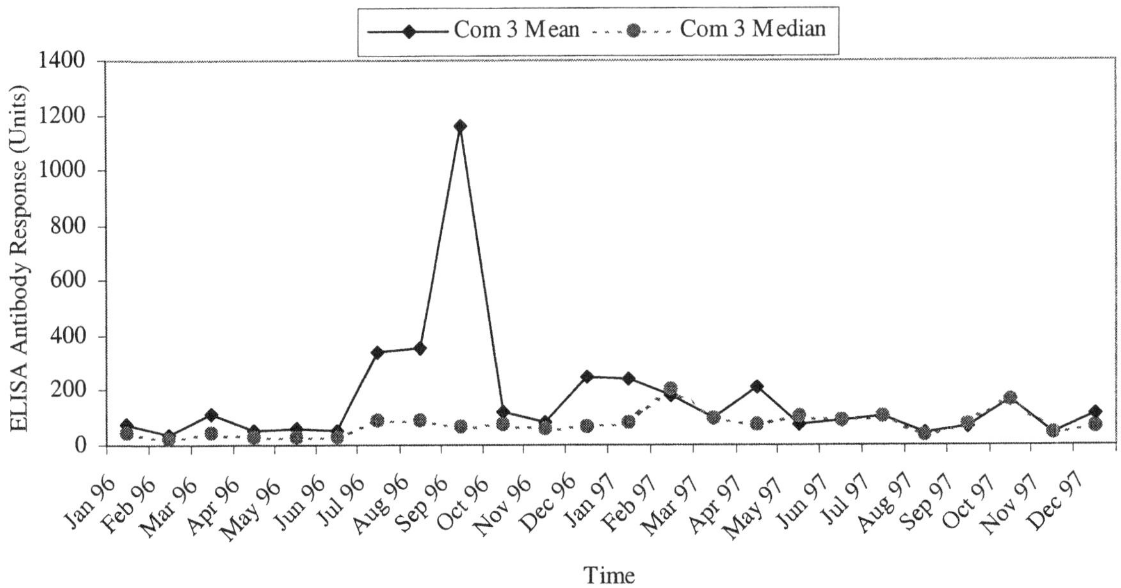

Figure 3.15 Community 3 - ELISA monthly mean and median for 17 kDa
(note differences in maximum y axis values between figures)

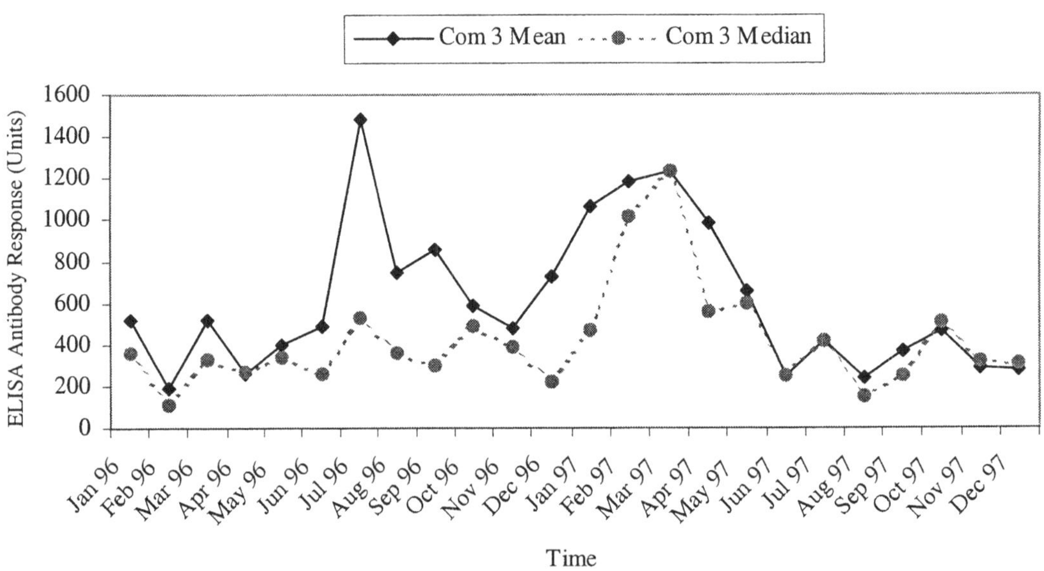

Figure 3.16 Community 3 - ELISA monthly mean and median for 27 kDa
(note differences in maximum y axis values between figures)

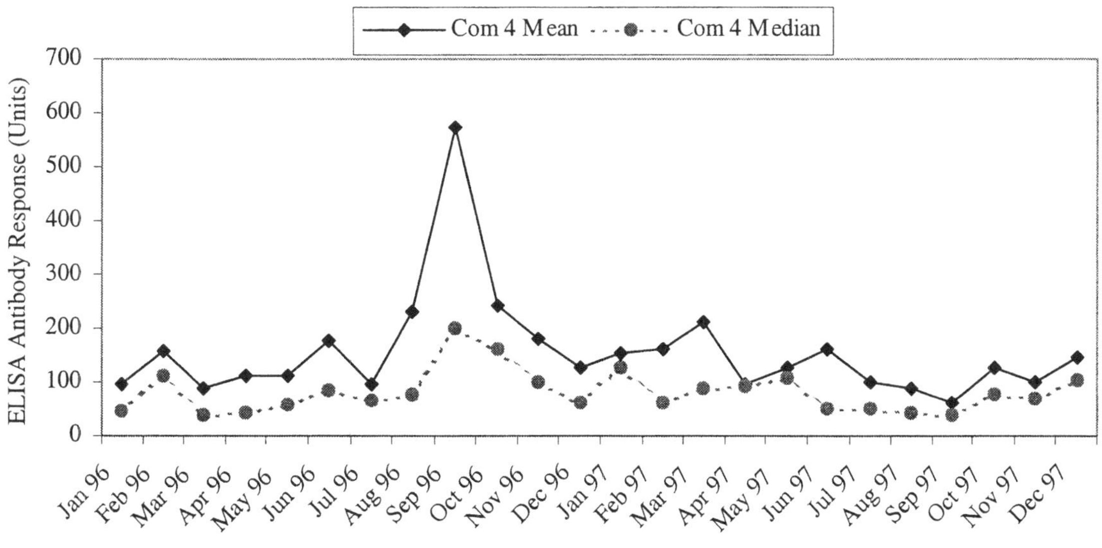

Figure 3.17 Community 4 - ELISA monthly mean and median for 17 kDa
(note differences in maximum y axis values between figures)

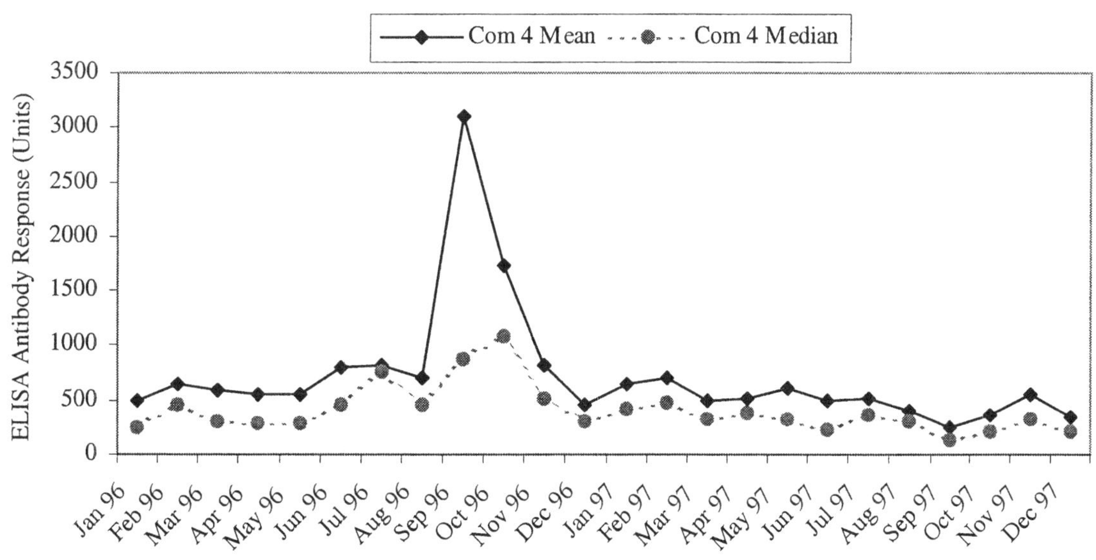

Figure 3.18 Community 4 - ELISA monthly mean and median for 27 kDa
(note differences in maximum y axis values between figures)

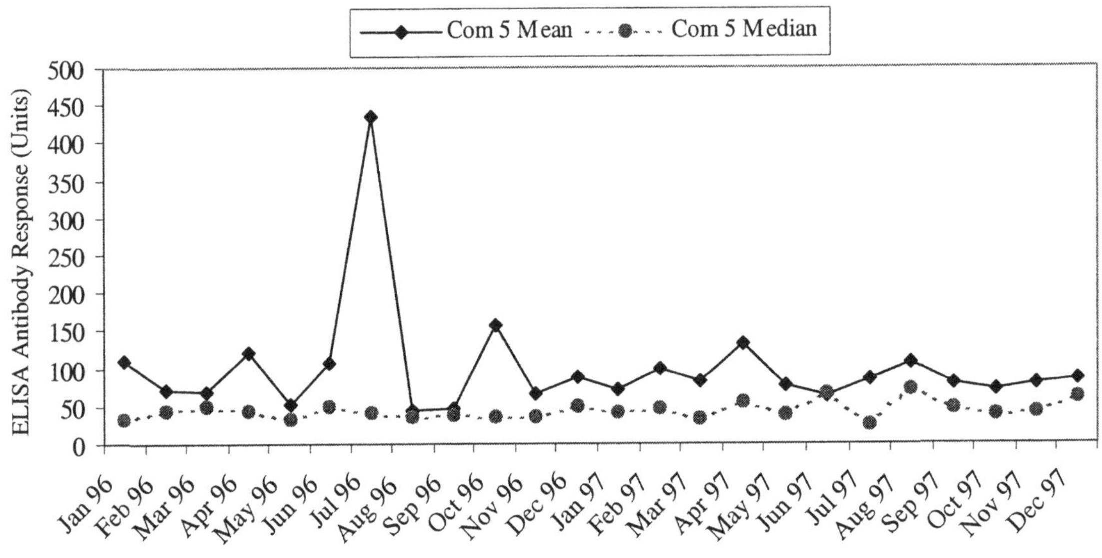

Figure 3.19 Community 5 - ELISA monthly mean and median for 17 kDa
(note differences in maximum y axis values between figures)

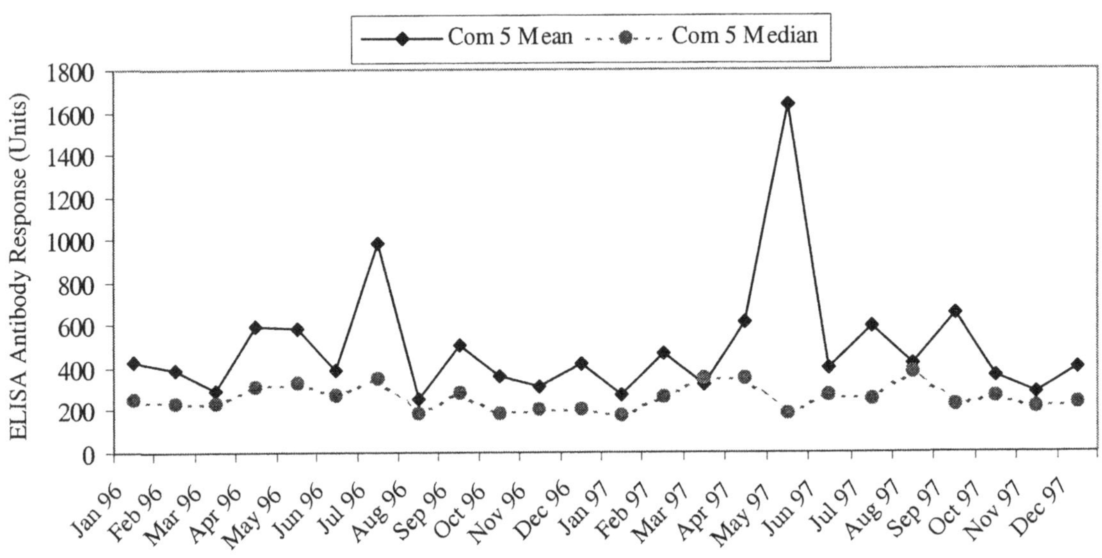

Figure 3.20 Community 5 - ELISA monthly mean and median for 27 kDa
(note differences in maximum y axis values between figures)

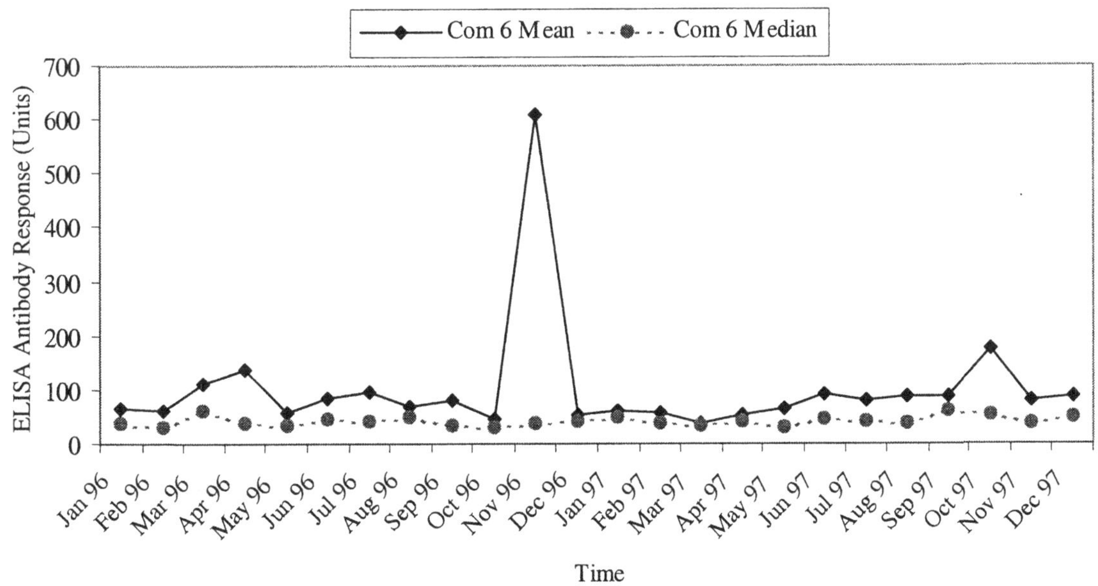

Figure 3.21 Community 6 - ELISA monthly mean and median for 17 kDa
(note differences in y axis values between figures)

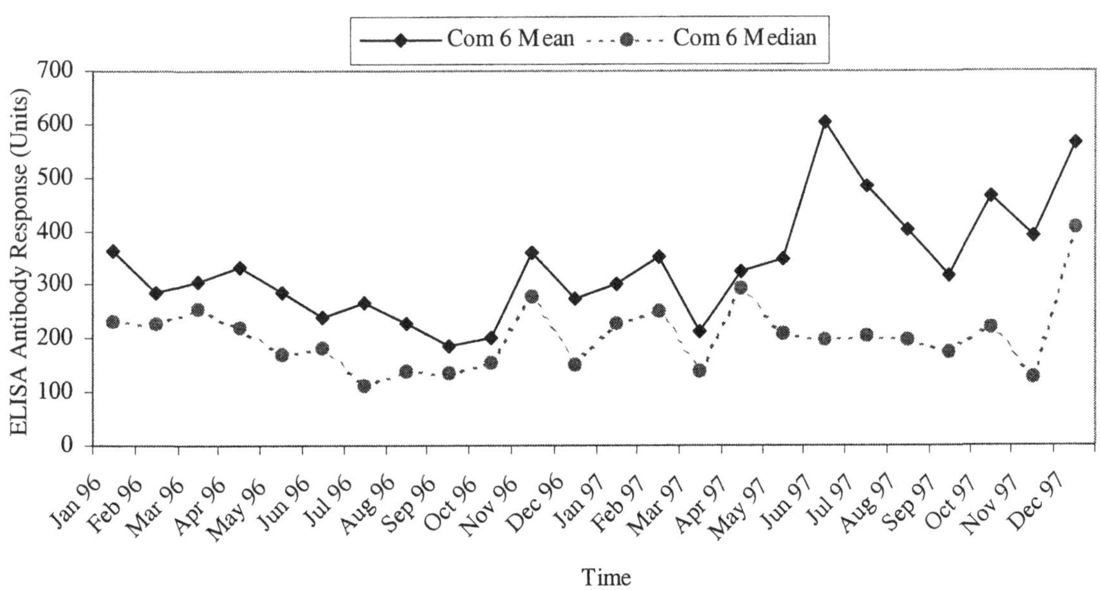

Figure 3.22 Community 6 - ELISA monthly mean and median for 27 kDa
(note differences in y axis values between figures)

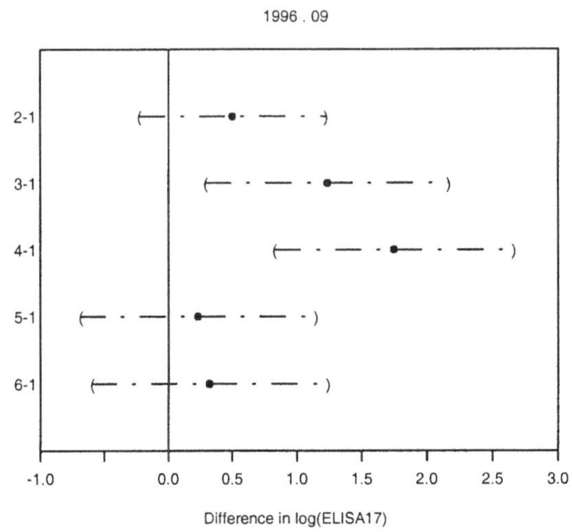

Figure 3.23 Simultaneous ELISA IgG 17 kDa comparisons of Community 2, 3, 4, 5, 6 to Community 1 for September 1996

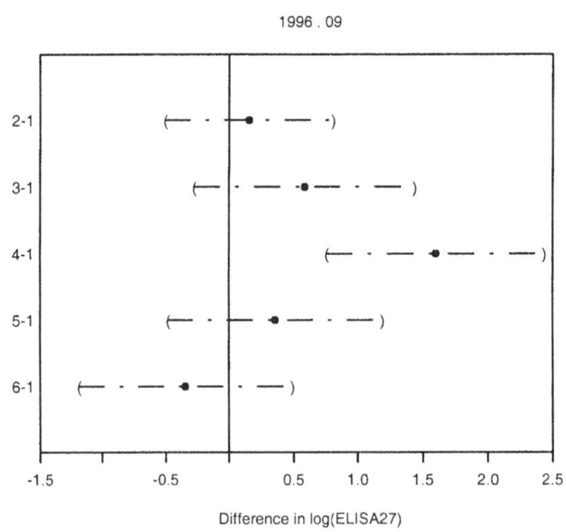

Figure 3.24 Simultaneous ELISA IgG 27 kDa comparisons of Community 2, 3, 4, 5, 6 to Community 1 for September 1996

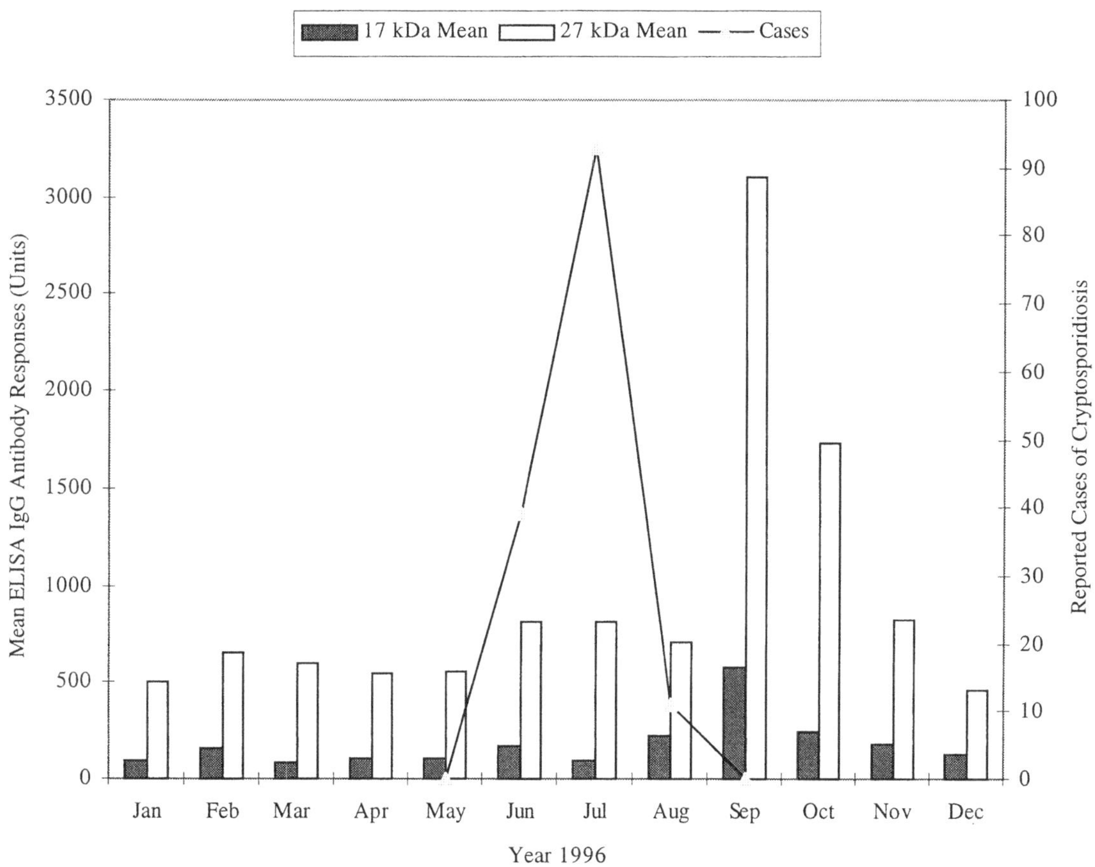

Figure 3.25 Comparison of monthly mean ELISA IgG antibody responses and monthly reported cases of cryptosporidiosis for Community 4 in 1996

CHAPTER 4
DISCUSSION

The limitations of a morphologically-based method for detection of *Cryptosporidium* in drinking water are well recognized (Clancy 2000). Craun *et al.* 1998 concluded that this method is not a useful predictor of waterborne outbreaks and therefore of community risk. Serological assays at a population level, a relatively new and promising tool, may be of use to public health in the study of epidemiological questions. The overall purpose of this study was to attempt to validate a practical serological method for use in this context. The project objectives were: to determine the sensitivity and host immune responses of *Cryptosporidium*-specific assays using unique sera from volunteers infected in waterborne outbreaks of cryptosporidiosis; to attempt to address some issues related to assay specificity; and to further assess the selected serological method at a population level.

An immunoblot method has been described. This test format however is laborious and not practical for the high throughput needed in large-scale epidemiological studies. When the Centres for Disease Control and Prevention (CDC) developed a second generation enzyme-linked immunosorbent assays (ELISA) for detection of antibodies to the 17 and 27 kDa immunodominant parasite antigens (Priest *et al.* 1999), this method was transferred to the British Columbia (UBC) laboratory and testing carried out to validate technology transfer. Comparison of results of the blinded testing of sera showed good correlation. The ELISA test format, a format which is used in most laboratories world-wide, was shown to be readily transferable from laboratory to laboratory.

Sera, collected from 67 cases of laboratory-confirmed cryptosporidiosis were tested by both immunoblot and ELISA assays. Results showed a higher sensitivity for the ELISA IgG 27 kDa assay (94%) when compared to the immunoblot 27 kDa (66%). Both immunoblot and ELISA 17 kDa assays showed similar and acceptable sensitivity (84% and 82% respectively). Both assays for circulating IgA showed unacceptably low levels of antibody against 17 and 27 kDa antigens (range 21-57%). The sensitivity of the ELISA assay for saliva specimens was also unacceptably low. This reflects at least in part, the nature of the samples being collected. High quality saliva specimens are difficult to obtain and it was clear that our specimens would not provide useful information. More success has been reported with oral fluid collectors which use a salt-impregnated pad to collect a transudate, a mixture of secreted and tissue antibodies. Studies using these collectors in outbreak settings have shown that the sensitivity of oral fluid ELISAs for *Cryptosporidium*-specific antibodies is approximately 70% relative to serum ELISAs (CDC, unpublished observations). Based on these findings, the ELISA IgG serum *Cryptosporidium* assays were chosen as the optimal format for further evaluation.

In discussions of the specificity of antibody tests for *Cryptosporidium*, it is important to recognise that this issue should be considered on two different levels. The specificity of a test such as the ELISA can be determined if we are using another test, e.g., the large format Western blot, as a gold standard. In this case, defining the performance (sensitivity and specificity) of the secondary test relative to the primary test is straightforward and easily accomplished. A more difficult task is defining the specificity of the antibody response. For many infectious diseases, the specificity of the antibody response is defined by analysing the antibody response of unexposed persons, those who have no contact with the organism. For *Cryptosporidium*, this approach is not possible because the parasite is ubiquitous and all human populations are probably exposed to infection. Thus, it may be impossible to identify a human population with no potential exposure to infection. Even neonates, by definition a naive population, will be born with maternal IgG.

In the absence of a truly unexposed population, we must use other strategies to determine whether infections with other organisms contribute to the antibody response to *C. parvum*. For example, if we collect specimens from persons infected with organism "X", we would expect to see changes in the antibody response to antigens derived from "X". If we do, indeed, see such changes, but observe no change in antibody response to *C. parvum*, this provides us with strong evidence that infection with "X" does not contribute to the antibody response to *C. parvum*. As parasites that may be waterborne, *Giardia* and *Toxoplasma*, are two important potential sources of cross reactivity. Serum specimens collected longitudinally from persons with *Toxoplasma* or *Giardia* show characteristic changes in antibody responses to antigens of the infecting agent but no change in antibody response to either the 17 or 27 kDa *C. parvum* antigens. For example, we observed that, in 5 paired sera which showed a rise in antibody levels to *Toxoplasma* antigens, there was no concomitant rise in IgG antibody to *Cryptosporidium*. This is the most clear-cut demonstration that these organisms do not contribute to the *Cryptosporidium*-specific antibody response; however, this approach may not be feasible for the entire universe of infections that may represent potential sources of cross-reactivity.

An alternative, if less rigorous, approach to the problem of cross reactivity and one that was used in this project was to test sera from persons with acute toxoplasmosis to conduct a quantitative comparison of antibody responses to *Toxoplasma* and *C. parvum*. If cross reactivity exists, we would expect to observe some correlation between antibody response to *Toxoplasma* and *C. parvum*, i.e. persons with high *Toxoplasma* antibody responses would be more likely to have higher *C. parvum* antibody responses. When sera from 40 cases of acute toxoplasmosis (some with low titres, some with intermediate and some with high anti-*Toxoplasma* titres) were tested for antibodies to *Cryptosporidium*, there was no correlation between levels of antibody against antigens of these two protozoans. This is consistent with the conclusion that there was no significant cross-reaction between these two parasites. A similar analysis was carried out on sera from persons acutely infected with *Giardia*. There was no positive correlation between levels of IgG antibody to *Cryptosporidium* and *Giardia*.

The pattern of host immune responses has been described previously using a large format immunoblot assay (Moss *et al.* 1994). We note that of the 109 stools collected (collection started within weeks of the initial diagnostic laboratory-positive stool specimens), only 1 stool from these laboratory-confirmed cases, was found to contain *Cryptosporidium* oocysts. This is consistent with the observation that oocyst shedding, at levels detectable by light microscopy, drops off relatively quickly after infection. On the other hand, anti-*Cryptosporidium* IgG antibody to the 17 kDa antigen remained above the cut-off value of 35 units for 51-60 weeks after infection and the IgG antibody to the 27 kDa antigen remained above the cut-off value of 86 units for an undefined length of time, minimally 100 weeks. While continued infections may not be detected by stool examination, and while obtaining fecal specimens for other than personal health-related causes is difficult, these limitations are widely understood by epidemiologists attempting to carry out population level studies. Parameters around the use of a serological approach must also be clearly defined and understood.

In this study, we examined the duration of the IgG antibody responses in the laboratory-confirmed cases after a waterborne exposure to *Cryptosporidium.* Some sera were drawn early enough as to demonstrate sero-conversion. Host response patterns varied from individual to individual, although all persons acquiring infection demonstrated a seropositivity that decreased over time. ELISA IgG to the 17 kDa antigen declined to below cut-off values at approximately 51-60 weeks post-onset. IgG response to the 27 kDa declined until approximately 51-60 weeks then remained at this level. This response did not fall below the cut-off value during the 100 weeks of the study. Since antibodies to the 27 kDa antigen may persist above the limit of detection, we suggest that this ELISA assay may be better suited to the assessment of historical exposure while the 17 kDa ELISA assay may be a better choice for the detection of recent infections (Priest *et al.* 2001). Longitudinal studies, however, that exceed 100 weeks are needed to determine if and when the 27 kDa antibody drops below the assay cut-off value.

During 1996, a waterborne outbreak occurred in Community 4. The number of cases reported to public health started to rise throughout May, peaking in July. Serum specimens drawn monthly from this population showed a rise and fall in the community levels of antibody to both 17 and 27 kDa antigens. This community's serological response lagged the epidemic curve by two months. Because of the intrinsic lag response, serology will probably not be a useful predictor of public health problems.

Analysis of community ELISA data showed significant differences across all communities. When these differences were explored further by comparing month-by-month results, effects of the Community 4 outbreak were clearly identifiable. Other serologically identifiable events, not related temporally to the known outbreaks, were also seen in several communities. For example, when antibody levels to the 17 kDa antigen of the population in Community 1 (deep well) were compared with that of the population in Community 2 (unfiltered surface water from protected watersheds), statistically significant differences were seen in January and March 1996, at least two months before the waterborne outbreaks started in the other communities. No waterborne outbreaks have been identified by public health in Community 2. Cryptosporidiosis has been a reportable communicable disease in British Columbia since 1996.

Both 17 and 27 kDa community antibody levels were significantly elevated during known waterborne outbreaks. The 17 kDa antibody levels however, appeared to detect more events in the surface water communities when compared with the deep well community. Results of the 27 kDa antibody testing showed less significant variability at the population level over time. The antibody to the 17 kDa antigen clearly reflected the effect of the largest known waterborne outbreak not only on the community (Community 4) in which it occurred but in the neighbouring community as well. These observations along with results of the Kinetic Cases testing, are consistent with these two antibody responses signifying two different responses at the community level.

Cryptosporidiosis is spread epidemically through exposures to contaminated food as well as water. Secondary transmission following an outbreak with ongoing person-to-person spread is also well documented. Since serology does not discriminate between modes of spread, we propose that within each community complex patterns of transmission are occurring. Further, we suggest that the serological trends indicate unidentified transmission events occurring within these communities.

Dynamic modelling using extended serological studies and their datasets need to be carried out. Given the basic data provided by the study of six communities for 24 months, further ELISA testing of the currently available but as yet untested community sera available for 1998, 1999 following this study (1996, 1997) should be carried out. A prospective serological study linked to active public health case detection or enhanced epidemiological surveillance could also expand our understanding of the events that were identified serologically in several communities.

CHAPTER 5
CONCLUSIONS AND RECOMMENDATIONS

The ELISA assay format is a practical, objective, quantifiable and reproducible method. It is a serological format that is readily transferable from one laboratory to another. The sensitivity of the ELISA assay and the immunoblot tests for IgG antibody to the 17 kDa antigen were both acceptable. In this study, the ELISA appeared to have greater sensitivity in the detection of IgG to the 27 kDa *Cryptosporidium* antigen. Since the ELISA format is less technically challenging than the immunoblot approach and can be readily used to test large numbers of sera required for epidemiological studies, it is the test format of choice for community studies.

Study of our cohort of laboratory-confirmed cases from a waterborne outbreak showed that both markers of infection declined over time with the 17 kDa declining the fastest to below the cutoff. The community serological response to both antibodies clearly identified the large confirmed waterborne outbreak of cryptosporidiosis (Community 4). Because the community serological response lagged many weeks behind the identification of the point source outbreaks, however, it is improbable that a serological approach will be a useful real-time public health tool. Both antibody markers detected the waterborne outbreaks although it appears that they measure different things. The 27 kDa ELISA assay may be better suited to the assessment of historical exposure while the 17 kDa may be a better choice for the detection of recent infections (Priest *et al.* 2001). At the present time, however, ELISA assays for both serological markers should be used.

Further community studies using these practical and sensitive ELISA assays are needed to enhance our understanding of the parameters around a serological approach to community studies. Testing of 1998 and 1999 sera which are now available from these six study communities should be carried out for use in the development of a mathematical model. This work is currently underway. It is noted that British Columbia has a traditional passive surveillance system in place for cryptosporidiosis. Further studies are underway to compare community serological activity with active laboratory surveillance. More information gathered from the use of these tools could be useful in a risk assessment model. In the event of another waterborne outbreak of cryptosporidiosis, it would be useful to collect sera from volunteers for a longer period of time to further study the declining 27 kDa response.

APPENDIX A

Table A.1

Data used for determination of 17 kDa cutoff

cutoff	tprate	fprate	tnrate	fnrate	tp	fp	fn	tn	propcorr
0	0.994	1.000	0.000	0.006	168	64	1	0	0.721
1	0.864	0.516	0.484	0.136	146	33	23	31	0.760
2	0.864	0.516	0.484	0.136	146	33	23	31	0.760
3	0.864	0.500	0.500	0.136	146	32	23	32	0.764
4	0.864	0.500	0.500	0.136	146	32	23	32	0.764
5	0.864	0.469	0.531	0.136	146	30	23	34	0.773
6	0.864	0.453	0.547	0.136	146	29	23	35	0.777
7	0.864	0.438	0.562	0.136	146	28	23	36	0.781
8	0.864	0.438	0.562	0.136	146	28	23	36	0.781
9	0.864	0.438	0.562	0.136	146	28	23	36	0.781
10	0.864	0.406	0.594	0.136	146	26	23	38	0.790
11	0.864	0.406	0.594	0.136	146	26	23	38	0.790
12	0.858	0.391	0.609	0.142	145	25	24	39	0.790
13	0.858	0.375	0.625	0.142	145	24	24	40	0.794
14	0.858	0.359	0.641	0.142	145	23	24	41	0.798
15	0.858	0.359	0.641	0.142	145	23	24	41	0.798
16	0.858	0.313	0.687	0.142	145	20	24	44	0.811
17	0.858	0.281	0.719	0.142	145	18	24	46	0.820
18	0.858	0.281	0.719	0.142	145	18	24	46	0.820
19	0.858	0.250	0.750	0.142	145	16	24	48	0.828
20	0.852	0.250	0.750	0.148	144	16	25	48	0.824
21	0.846	0.219	0.781	0.154	143	14	26	50	0.828
22	0.846	0.219	0.781	0.154	143	14	26	50	0.828
23	0.840	0.219	0.781	0.160	142	14	27	50	0.824
24	0.822	0.172	0.828	0.178	139	11	30	53	0.824
25	0.817	0.172	0.828	0.183	138	11	31	53	0.820

Table A.1 (continued)

cutoff	tprate	fprate	tnrate	fnrate	tp	fp	fn	tn	propcorr
26	0.805	0.172	0.828	0.195	136	11	33	53	0.811
27	0.799	0.172	0.828	0.201	135	11	34	53	0.807
28	0.799	0.172	0.828	0.201	135	11	34	53	0.807
29	0.799	0.172	0.828	0.201	135	11	34	53	0.807
30	0.793	0.156	0.844	0.207	134	10	35	54	0.807
31	0.793	0.141	0.859	0.207	134	9	35	55	0.811
32	0.793	0.141	0.859	0.207	134	9	35	55	0.811
33	0.793	0.141	0.859	0.207	134	9	35	55	0.811
34	0.793	0.141	0.859	0.207	134	9	35	55	0.811
35	0.793	0.141	0.859	0.207	134	9	35	55	0.811
36	0.793	0.141	0.859	0.207	134	9	35	55	0.811
37	0.781	0.141	0.859	0.219	132	9	37	55	0.803
38	0.775	0.141	0.859	0.225	131	9	38	55	0.798
39	0.775	0.141	0.859	0.225	131	9	38	55	0.798
40	0.769	0.141	0.859	0.231	130	9	39	55	0.794
41	0.769	0.141	0.859	0.231	130	9	39	55	0.794
42	0.763	0.141	0.859	0.237	129	9	40	55	0.790
43	0.751	0.141	0.859	0.249	127	9	42	55	0.781
44	0.740	0.141	0.859	0.260	125	9	44	55	0.773
45	0.734	0.141	0.859	0.266	124	9	45	55	0.768
46	0.728	0.141	0.859	0.272	123	9	46	55	0.764
47	0.728	0.141	0.859	0.272	123	9	46	55	0.764
48	0.728	0.141	0.859	0.272	123	9	46	55	0.764
49	0.710	0.141	0.859	0.290	120	9	49	55	0.751
50	0.704	0.141	0.859	0.296	119	9	50	55	0.747
51	0.704	0.141	0.859	0.296	119	9	50	55	0.747
52	0.704	0.141	0.859	0.296	119	9	50	55	0.747

Table A.1 (continued)

cutoff	tprate	fprate	tnrate	fnrate	tp	fp	fn	tn	propcorr
53	0.686	0.141	0.859	0.314	116	9	53	55	0.734
54	0.686	0.141	0.859	0.314	116	9	53	55	0.734
55	0.675	0.125	0.875	0.325	114	8	55	56	0.730
56	0.669	0.125	0.875	0.331	113	8	56	56	0.725
57	0.663	0.125	0.875	0.337	112	8	57	56	0.721
58	0.645	0.125	0.875	0.355	109	8	60	56	0.708
59	0.627	0.125	0.875	0.373	106	8	63	56	0.695
60	0.615	0.125	0.875	0.385	104	8	65	56	0.687
61	0.586	0.125	0.875	0.414	99	8	70	56	0.665
62	0.580	0.125	0.875	0.420	98	8	71	56	0.661
63	0.574	0.125	0.875	0.426	97	8	72	56	0.657
64	0.574	0.125	0.875	0.426	97	8	72	56	0.657
65	0.574	0.125	0.875	0.426	97	8	72	56	0.657
66	0.562	0.125	0.875	0.438	95	8	74	56	0.648
67	0.562	0.125	0.875	0.438	95	8	74	56	0.648
68	0.562	0.125	0.875	0.438	95	8	74	56	0.648
69	0.556	0.125	0.875	0.444	94	8	75	56	0.644
70	0.550	0.125	0.875	0.450	93	8	76	56	0.639
71	0.550	0.125	0.875	0.450	93	8	76	56	0.639
72	0.544	0.125	0.875	0.456	92	8	77	56	0.635
73	0.544	0.125	0.875	0.456	92	8	77	56	0.635
74	0.538	0.125	0.875	0.462	91	8	78	56	0.631
75	0.533	0.125	0.875	0.467	90	8	79	56	0.627
76	0.533	0.125	0.875	0.467	90	8	79	56	0.627
77	0.533	0.125	0.875	0.467	90	8	79	56	0.627
78	0.533	0.125	0.875	0.467	90	8	79	56	0.627
79	0.533	0.125	0.875	0.467	90	8	79	56	0.627

Table A.1 (continued)

cutoff	tprate	fprate	tnrate	fnrate	tp	fp	fn	tn	propcorr
80	0.533	0.125	0.875	0.467	90	8	79	56	0.627
81	0.533	0.125	0.875	0.467	90	8	79	56	0.627
82	0.533	0.125	0.875	0.467	90	8	79	56	0.627
83	0.533	0.125	0.875	0.467	90	8	79	56	0.627
84	0.533	0.125	0.875	0.467	90	8	79	56	0.627
85	0.533	0.109	0.891	0.467	90	7	79	57	0.631
86	0.533	0.109	0.891	0.467	90	7	79	57	0.631
87	0.533	0.109	0.891	0.467	90	7	79	57	0.631
88	0.527	0.109	0.891	0.473	89	7	80	57	0.627
89	0.521	0.109	0.891	0.479	88	7	81	57	0.622
90	0.521	0.109	0.891	0.479	88	7	81	57	0.622
91	0.521	0.109	0.891	0.479	88	7	81	57	0.622
92	0.521	0.109	0.891	0.479	88	7	81	57	0.622
93	0.521	0.109	0.891	0.479	88	7	81	57	0.622
94	0.515	0.109	0.891	0.485	87	7	82	57	0.618
95	0.515	0.109	0.891	0.485	87	7	82	57	0.618
96	0.515	0.109	0.891	0.485	87	7	82	57	0.618
97	0.515	0.109	0.891	0.485	87	7	82	57	0.618
98	0.515	0.109	0.891	0.485	87	7	82	57	0.618
99	0.515	0.109	0.891	0.485	87	7	82	57	0.618
100	0.497	0.109	0.891	0.503	84	7	85	57	0.605
101	0.491	0.109	0.891	0.509	83	7	86	57	0.601
102	0.491	0.109	0.891	0.509	83	7	86	57	0.601
103	0.485	0.109	0.891	0.515	82	7	87	57	0.597
104	0.485	0.109	0.891	0.515	82	7	87	57	0.597
105	0.479	0.109	0.891	0.521	81	7	88	57	0.592
106	0.479	0.109	0.891	0.521	81	7	88	57	0.592

Table A.1 (continued)

cutoff	tprate	fprate	tnrate	fnrate	tp	fp	fn	tn	propcorr
107	0.473	0.109	0.891	0.527	80	7	89	57	0.588
108	0.467	0.109	0.891	0.533	79	7	90	57	0.584
109	0.467	0.109	0.891	0.533	79	7	90	57	0.584
110	0.467	0.109	0.891	0.533	79	7	90	57	0.584
111	0.462	0.109	0.891	0.538	78	7	91	57	0.579
112	0.462	0.109	0.891	0.538	78	7	91	57	0.579
113	0.456	0.109	0.891	0.544	77	7	92	57	0.575
114	0.456	0.109	0.891	0.544	77	7	92	57	0.575
115	0.450	0.109	0.891	0.550	76	7	93	57	0.571
116	0.444	0.109	0.891	0.556	75	7	94	57	0.567
117	0.444	0.109	0.891	0.556	75	7	94	57	0.567
118	0.444	0.109	0.891	0.556	75	7	94	57	0.567
119	0.444	0.109	0.891	0.556	75	7	94	57	0.567
120	0.444	0.109	0.891	0.556	75	7	94	57	0.567
121	0.444	0.109	0.891	0.556	75	7	94	57	0.567
122	0.438	0.109	0.891	0.562	74	7	95	57	0.562
123	0.438	0.109	0.891	0.562	74	7	95	57	0.562
124	0.438	0.109	0.891	0.562	74	7	95	57	0.562
125	0.438	0.109	0.891	0.562	74	7	95	57	0.562
126	0.438	0.109	0.891	0.562	74	7	95	57	0.562
127	0.438	0.109	0.891	0.562	74	7	95	57	0.562
128	0.438	0.109	0.891	0.562	74	7	95	57	0.562
129	0.438	0.109	0.891	0.562	74	7	95	57	0.562
130	0.438	0.109	0.891	0.562	74	7	95	57	0.562
131	0.438	0.109	0.891	0.562	74	7	95	57	0.562
132	0.438	0.109	0.891	0.562	74	7	95	57	0.562
133	0.438	0.109	0.891	0.562	74	7	95	57	0.562

Table A.1 (continued)

cutoff	tprate	fprate	tnrate	fnrate	tp	fp	fn	tn	propcorr
134	0.438	0.109	0.891	0.562	74	7	95	57	0.562
135	0.438	0.109	0.891	0.562	74	7	95	57	0.562
136	0.438	0.109	0.891	0.562	74	7	95	57	0.562
137	0.432	0.109	0.891	0.568	73	7	96	57	0.558
138	0.432	0.109	0.891	0.568	73	7	96	57	0.558
139	0.426	0.109	0.891	0.574	72	7	97	57	0.554
140	0.420	0.109	0.891	0.580	71	7	98	57	0.549
141	0.414	0.109	0.891	0.586	70	7	99	57	0.545
142	0.408	0.109	0.891	0.592	69	7	100	57	0.541
143	0.408	0.109	0.891	0.592	69	7	100	57	0.541
144	0.408	0.109	0.891	0.592	69	7	100	57	0.541
145	0.408	0.109	0.891	0.592	69	7	100	57	0.541
146	0.408	0.109	0.891	0.592	69	7	100	57	0.541
147	0.402	0.109	0.891	0.598	68	7	101	57	0.536
148	0.402	0.109	0.891	0.598	68	7	101	57	0.536
149	0.402	0.109	0.891	0.598	68	7	101	57	0.536
150	0.396	0.109	0.891	0.604	67	7	102	57	0.532
151	0.396	0.109	0.891	0.604	67	7	102	57	0.532
152	0.396	0.109	0.891	0.604	67	7	102	57	0.532
153	0.396	0.109	0.891	0.604	67	7	102	57	0.532
154	0.396	0.109	0.891	0.604	67	7	102	57	0.532
155	0.396	0.109	0.891	0.604	67	7	102	57	0.532
156	0.396	0.109	0.891	0.604	67	7	102	57	0.532
157	0.396	0.109	0.891	0.604	67	7	102	57	0.532
158	0.396	0.109	0.891	0.604	67	7	102	57	0.532
159	0.396	0.109	0.891	0.604	67	7	102	57	0.532
160	0.396	0.109	0.891	0.604	67	7	102	57	0.532

Table A.1 (continued)

cutoff	tprate	fprate	tnrate	fnrate	tp	fp	fn	tn	propcorr
161	0.396	0.109	0.891	0.604	67	7	102	57	0.532
162	0.396	0.109	0.891	0.604	67	7	102	57	0.532
163	0.396	0.109	0.891	0.604	67	7	102	57	0.532
164	0.396	0.109	0.891	0.604	67	7	102	57	0.532
165	0.396	0.109	0.891	0.604	67	7	102	57	0.532
166	0.396	0.109	0.891	0.604	67	7	102	57	0.532
167	0.391	0.109	0.891	0.609	66	7	103	57	0.528
168	0.391	0.109	0.891	0.609	66	7	103	57	0.528
169	0.391	0.109	0.891	0.609	66	7	103	57	0.528
170	0.391	0.109	0.891	0.609	66	7	103	57	0.528
171	0.391	0.109	0.891	0.609	66	7	103	57	0.528
172	0.391	0.109	0.891	0.609	66	7	103	57	0.528
173	0.391	0.109	0.891	0.609	66	7	103	57	0.528
174	0.391	0.109	0.891	0.609	66	7	103	57	0.528
175	0.391	0.109	0.891	0.609	66	7	103	57	0.528
176	0.391	0.109	0.891	0.609	66	7	103	57	0.528
177	0.385	0.109	0.891	0.615	65	7	104	57	0.524
178	0.385	0.109	0.891	0.615	65	7	104	57	0.524
179	0.385	0.094	0.906	0.615	65	6	104	58	0.528
180	0.385	0.094	0.906	0.615	65	6	104	58	0.528
181	0.385	0.094	0.906	0.615	65	6	104	58	0.528
182	0.385	0.094	0.906	0.615	65	6	104	58	0.528
183	0.379	0.094	0.906	0.621	64	6	105	58	0.524
184	0.379	0.094	0.906	0.621	64	6	105	58	0.524
185	0.379	0.094	0.906	0.621	64	6	105	58	0.524
186	0.379	0.094	0.906	0.621	64	6	105	58	0.524
187	0.379	0.094	0.906	0.621	64	6	105	58	0.524

Table A.1 (continued)

cutoff	tprate	fprate	tnrate	fnrate	tp	fp	fn	tn	propcorr
188	0.379	0.094	0.906	0.621	64	6	105	58	0.524
189	0.379	0.094	0.906	0.621	64	6	105	58	0.524
190	0.373	0.094	0.906	0.627	63	6	106	58	0.519
191	0.373	0.094	0.906	0.627	63	6	106	58	0.519
192	0.373	0.094	0.906	0.627	63	6	106	58	0.519
193	0.373	0.094	0.906	0.627	63	6	106	58	0.519
194	0.373	0.094	0.906	0.627	63	6	106	58	0.519
195	0.373	0.094	0.906	0.627	63	6	106	58	0.519
196	0.373	0.094	0.906	0.627	63	6	106	58	0.519
197	0.373	0.094	0.906	0.627	63	6	106	58	0.519
198	0.373	0.094	0.906	0.627	63	6	106	58	0.519
199	0.367	0.094	0.906	0.633	62	6	107	58	0.515
200	0.367	0.094	0.906	0.633	62	6	107	58	0.515
201	0.367	0.094	0.906	0.633	62	6	107	58	0.515
202	0.361	0.094	0.906	0.639	61	6	108	58	0.511
203	0.361	0.094	0.906	0.639	61	6	108	58	0.511
204	0.355	0.094	0.906	0.645	60	6	109	58	0.506
205	0.355	0.094	0.906	0.645	60	6	109	58	0.506
206	0.355	0.094	0.906	0.645	60	6	109	58	0.506
207	0.343	0.094	0.906	0.657	58	6	111	58	0.498
208	0.343	0.094	0.906	0.657	58	6	111	58	0.498
209	0.343	0.094	0.906	0.657	58	6	111	58	0.498
210	0.343	0.094	0.906	0.657	58	6	111	58	0.498
211	0.343	0.094	0.906	0.657	58	6	111	58	0.498
212	0.343	0.094	0.906	0.657	58	6	111	58	0.498
213	0.337	0.094	0.906	0.663	57	6	112	58	0.494
214	0.337	0.094	0.906	0.663	57	6	112	58	0.494

Table A.1 (continued)

cutoff	tprate	fprate	tnrate	fnrate	tp	fp	fn	tn	propcorr
215	0.337	0.094	0.906	0.663	57	6	112	58	0.494
216	0.337	0.094	0.906	0.663	57	6	112	58	0.494
217	0.337	0.094	0.906	0.663	57	6	112	58	0.494
218	0.337	0.094	0.906	0.663	57	6	112	58	0.494
219	0.331	0.094	0.906	0.669	56	6	113	58	0.489
220	0.325	0.094	0.906	0.675	55	6	114	58	0.485
221	0.325	0.094	0.906	0.675	55	6	114	58	0.485
222	0.325	0.094	0.906	0.675	55	6	114	58	0.485
223	0.325	0.094	0.906	0.675	55	6	114	58	0.485
224	0.325	0.094	0.906	0.675	55	6	114	58	0.485
225	0.325	0.094	0.906	0.675	55	6	114	58	0.485
226	0.320	0.094	0.906	0.680	54	6	115	58	0.481
227	0.320	0.078	0.922	0.680	54	5	115	59	0.485
228	0.320	0.078	0.922	0.680	54	5	115	59	0.485
229	0.320	0.078	0.922	0.680	54	5	115	59	0.485
230	0.314	0.078	0.922	0.686	53	5	116	59	0.481
231	0.308	0.078	0.922	0.692	52	5	117	59	0.476
232	0.308	0.078	0.922	0.692	52	5	117	59	0.476
233	0.302	0.078	0.922	0.698	51	5	118	59	0.472
234	0.302	0.078	0.922	0.698	51	5	118	59	0.472
235	0.296	0.078	0.922	0.704	50	5	119	59	0.468
236	0.296	0.078	0.922	0.704	50	5	119	59	0.468
237	0.296	0.078	0.922	0.704	50	5	119	59	0.468
238	0.296	0.078	0.922	0.704	50	5	119	59	0.468
239	0.296	0.078	0.922	0.704	50	5	119	59	0.468
240	0.290	0.078	0.922	0.710	49	5	120	59	0.464
241	0.290	0.078	0.922	0.710	49	5	120	59	0.464

Table A.1 (continued)

cutoff	tprate	fprate	tnrate	fnrate	tp	fp	fn	tn	propcorr
242	0.290	0.078	0.922	0.710	49	5	120	59	0.464
243	0.290	0.078	0.922	0.710	49	5	120	59	0.464
244	0.290	0.078	0.922	0.710	49	5	120	59	0.464
245	0.290	0.063	0.937	0.710	49	4	120	60	0.468
246	0.290	0.063	0.937	0.710	49	4	120	60	0.468
247	0.290	0.063	0.937	0.710	49	4	120	60	0.468
248	0.290	0.063	0.937	0.710	49	4	120	60	0.468
249	0.284	0.063	0.937	0.716	48	4	121	60	0.464
250	0.278	0.047	0.953	0.722	47	3	122	61	0.464
251	0.278	0.047	0.953	0.722	47	3	122	61	0.464
252	0.278	0.047	0.953	0.722	47	3	122	61	0.464
253	0.278	0.047	0.953	0.722	47	3	122	61	0.464
254	0.278	0.047	0.953	0.722	47	3	122	61	0.464
255	0.278	0.047	0.953	0.722	47	3	122	61	0.464
256	0.278	0.047	0.953	0.722	47	3	122	61	0.464
257	0.278	0.047	0.953	0.722	47	3	122	61	0.464
258	0.272	0.047	0.953	0.728	46	3	123	61	0.459
259	0.272	0.031	0.969	0.728	46	2	123	62	0.464
260	0.272	0.031	0.969	0.728	46	2	123	62	0.464
261	0.272	0.031	0.969	0.728	46	2	123	62	0.464
262	0.272	0.031	0.969	0.728	46	2	123	62	0.464
263	0.272	0.031	0.969	0.728	46	2	123	62	0.464
264	0.272	0.031	0.969	0.728	46	2	123	62	0.464
265	0.272	0.031	0.969	0.728	46	2	123	62	0.464
266	0.272	0.031	0.969	0.728	46	2	123	62	0.464
267	0.272	0.031	0.969	0.728	46	2	123	62	0.464
268	0.272	0.031	0.969	0.728	46	2	123	62	0.464

Table A.1 (continued)

cutoff	tprate	fprate	tnrate	fnrate	tp	fp	fn	tn	propcorr
269	0.272	0.031	0.969	0.728	46	2	123	62	0.464
270	0.272	0.031	0.969	0.728	46	2	123	62	0.464
271	0.272	0.031	0.969	0.728	46	2	123	62	0.464
272	0.266	0.031	0.969	0.734	45	2	124	62	0.459
273	0.260	0.031	0.969	0.740	44	2	125	62	0.455
274	0.260	0.031	0.969	0.740	44	2	125	62	0.455
275	0.260	0.031	0.969	0.740	44	2	125	62	0.455
276	0.249	0.031	0.969	0.751	42	2	127	62	0.446
277	0.249	0.031	0.969	0.751	42	2	127	62	0.446
278	0.249	0.031	0.969	0.751	42	2	127	62	0.446
279	0.249	0.031	0.969	0.751	42	2	127	62	0.446
280	0.249	0.031	0.969	0.751	42	2	127	62	0.446
281	0.243	0.031	0.969	0.757	41	2	128	62	0.442
282	0.243	0.031	0.969	0.757	41	2	128	62	0.442
283	0.243	0.031	0.969	0.757	41	2	128	62	0.442
284	0.243	0.031	0.969	0.757	41	2	128	62	0.442
285	0.243	0.031	0.969	0.757	41	2	128	62	0.442
286	0.243	0.031	0.969	0.757	41	2	128	62	0.442
287	0.243	0.031	0.969	0.757	41	2	128	62	0.442
288	0.243	0.031	0.969	0.757	41	2	128	62	0.442
289	0.243	0.031	0.969	0.757	41	2	128	62	0.442
290	0.243	0.031	0.969	0.757	41	2	128	62	0.442
291	0.243	0.031	0.969	0.757	41	2	128	62	0.442
292	0.243	0.031	0.969	0.757	41	2	128	62	0.442
293	0.237	0.031	0.969	0.763	40	2	129	62	0.438
294	0.237	0.031	0.969	0.763	40	2	129	62	0.438
295	0.237	0.031	0.969	0.763	40	2	129	62	0.438

Table A.1 (continued)

cutoff	tprate	fprate	tnrate	fnrate	tp	fp	fn	tn	propcorr
296	0.237	0.031	0.969	0.763	40	2	129	62	0.438
297	0.237	0.031	0.969	0.763	40	2	129	62	0.438
298	0.237	0.031	0.969	0.763	40	2	129	62	0.438
299	0.237	0.031	0.969	0.763	40	2	129	62	0.438
300	0.237	0.031	0.969	0.763	40	2	129	62	0.438

Table A.2

Data used for determination of 27 kDa cutoff

cutoff	tprate	fprate	tnrate	fnrate	tp	fp	fn	tn	propcorr
0	0.995	1.000	0.000	0.005	217	15	1	0	0.931
1	0.963	1.000	0.000	0.037	210	15	8	0	0.901
2	0.963	1.000	0.000	0.037	210	15	8	0	0.901
3	0.963	1.000	0.000	0.037	210	15	8	0	0.901
4	0.963	1.000	0.000	0.037	210	15	8	0	0.901
5	0.963	1.000	0.000	0.037	210	15	8	0	0.901
6	0.963	1.000	0.000	0.037	210	15	8	0	0.901
7	0.963	1.000	0.000	0.037	210	15	8	0	0.901
8	0.963	1.000	0.000	0.037	210	15	8	0	0.901
9	0.963	1.000	0.000	0.037	210	15	8	0	0.901
10	0.963	1.000	0.000	0.037	210	15	8	0	0.901
11	0.963	1.000	0.000	0.037	210	15	8	0	0.901
12	0.963	1.000	0.000	0.037	210	15	8	0	0.901
13	0.963	1.000	0.000	0.037	210	15	8	0	0.901
14	0.963	1.000	0.000	0.037	210	15	8	0	0.901
15	0.963	1.000	0.000	0.037	210	15	8	0	0.901
16	0.963	1.000	0.000	0.037	210	15	8	0	0.901
17	0.963	1.000	0.000	0.037	210	15	8	0	0.901
18	0.963	1.000	0.000	0.037	210	15	8	0	0.901
19	0.963	1.000	0.000	0.037	210	15	8	0	0.901
20	0.963	1.000	0.000	0.037	210	15	8	0	0.901
21	0.963	1.000	0.000	0.037	210	15	8	0	0.901
22	0.963	1.000	0.000	0.037	210	15	8	0	0.901
23	0.963	1.000	0.000	0.037	210	15	8	0	0.901
24	0.963	1.000	0.000	0.037	210	15	8	0	0.901
25	0.963	1.000	0.000	0.037	210	15	8	0	0.901

Table A.2 (continued)

cutoff	tprate	fprate	tnrate	fnrate	tp	fp	fn	tn	propcorr
26	0.963	1.000	0.000	0.037	210	15	8	0	0.901
27	0.963	1.000	0.000	0.037	210	15	8	0	0.901
28	0.963	1.000	0.000	0.037	210	15	8	0	0.901
29	0.963	1.000	0.000	0.037	210	15	8	0	0.901
30	0.963	1.000	0.000	0.037	210	15	8	0	0.901
31	0.963	1.000	0.000	0.037	210	15	8	0	0.901
32	0.959	1.000	0.000	0.041	209	15	9	0	0.897
33	0.959	1.000	0.000	0.041	209	15	9	0	0.897
34	0.959	1.000	0.000	0.041	209	15	9	0	0.897
35	0.959	1.000	0.000	0.041	209	15	9	0	0.897
36	0.959	1.000	0.000	0.041	209	15	9	0	0.897
37	0.959	1.000	0.000	0.041	209	15	9	0	0.897
38	0.959	1.000	0.000	0.041	209	15	9	0	0.897
39	0.959	1.000	0.000	0.041	209	15	9	0	0.897
40	0.954	1.000	0.000	0.046	208	15	10	0	0.893
41	0.950	1.000	0.000	0.050	207	15	11	0	0.888
42	0.950	0.867	0.133	0.050	207	13	11	2	0.897
43	0.950	0.867	0.133	0.050	207	13	11	2	0.897
44	0.950	0.867	0.133	0.050	207	13	11	2	0.897
45	0.950	0.733	0.267	0.050	207	11	11	4	0.906
46	0.950	0.667	0.333	0.050	207	10	11	5	0.910
47	0.950	0.667	0.333	0.050	207	10	11	5	0.910
48	0.945	0.667	0.333	0.055	206	10	12	5	0.906
49	0.945	0.667	0.333	0.055	206	10	12	5	0.906
50	0.945	0.533	0.467	0.055	206	8	12	7	0.914
51	0.940	0.533	0.467	0.060	205	8	13	7	0.910
52	0.940	0.533	0.467	0.060	205	8	13	7	0.910

Table A.2 (continued)

cutoff	tprate	fprate	tnrate	fnrate	tp	fp	fn	tn	propcorr
53	0.940	0.533	0.467	0.060	205	8	13	7	0.910
54	0.940	0.533	0.467	0.060	205	8	13	7	0.910
55	0.940	0.533	0.467	0.060	205	8	13	7	0.910
56	0.940	0.400	0.600	0.060	205	6	13	9	0.918
57	0.940	0.400	0.600	0.060	205	6	13	9	0.918
58	0.940	0.400	0.600	0.060	205	6	13	9	0.918
59	0.940	0.400	0.600	0.060	205	6	13	9	0.918
60	0.940	0.400	0.600	0.060	205	6	13	9	0.918
61	0.940	0.333	0.667	0.060	205	5	13	10	0.923
62	0.940	0.333	0.667	0.060	205	5	13	10	0.923
63	0.936	0.267	0.733	0.064	204	4	14	11	0.923
64	0.936	0.267	0.733	0.064	204	4	14	11	0.923
65	0.931	0.267	0.733	0.069	203	4	15	11	0.918
66	0.927	0.200	0.800	0.073	202	3	16	12	0.918
67	0.927	0.200	0.800	0.073	202	3	16	12	0.918
68	0.922	0.200	0.800	0.078	201	3	17	12	0.914
69	0.922	0.067	0.933	0.078	201	1	17	14	0.923
70	0.922	0.067	0.933	0.078	201	1	17	14	0.923
71	0.922	0.067	0.933	0.078	201	1	17	14	0.923
72	0.913	0.067	0.933	0.087	199	1	19	14	0.914
73	0.908	0.067	0.933	0.092	198	1	20	14	0.910
74	0.904	0.000	1.000	0.096	197	0	21	15	0.910
75	0.904	0.000	1.000	0.096	197	0	21	15	0.910
76	0.890	0.000	1.000	0.110	194	0	24	15	0.897
77	0.885	0.000	1.000	0.115	193	0	25	15	0.893
78	0.885	0.000	1.000	0.115	193	0	25	15	0.893
79	0.885	0.000	1.000	0.115	193	0	25	15	0.893

Table A.2 (continued)

cutoff	tprate	fprate	tnrate	fnrate	tp	fp	fn	tn	propcorr
80	0.885	0.000	1.000	0.115	193	0	25	15	0.893
81	0.881	0.000	1.000	0.119	192	0	26	15	0.888
82	0.881	0.000	1.000	0.119	192	0	26	15	0.888
83	0.881	0.000	1.000	0.119	192	0	26	15	0.888
84	0.867	0.000	1.000	0.133	189	0	29	15	0.876
85	0.867	0.000	1.000	0.133	189	0	29	15	0.876
86	0.867	0.000	1.000	0.133	189	0	29	15	0.876
87	0.867	0.000	1.000	0.133	189	0	29	15	0.876
88	0.867	0.000	1.000	0.133	189	0	29	15	0.876
89	0.862	0.000	1.000	0.138	188	0	30	15	0.871
90	0.862	0.000	1.000	0.138	188	0	30	15	0.871
91	0.862	0.000	1.000	0.138	188	0	30	15	0.871
92	0.858	0.000	1.000	0.142	187	0	31	15	0.867
93	0.853	0.000	1.000	0.147	186	0	32	15	0.863
94	0.849	0.000	1.000	0.151	185	0	33	15	0.858
95	0.849	0.000	1.000	0.151	185	0	33	15	0.858
96	0.844	0.000	1.000	0.156	184	0	34	15	0.854
97	0.844	0.000	1.000	0.156	184	0	34	15	0.854
98	0.835	0.000	1.000	0.165	182	0	36	15	0.845
99	0.830	0.000	1.000	0.170	181	0	37	15	0.841
100	0.830	0.000	1.000	0.170	181	0	37	15	0.841
101	0.821	0.000	1.000	0.179	179	0	39	15	0.833
102	0.817	0.000	1.000	0.183	178	0	40	15	0.828
103	0.817	0.000	1.000	0.183	178	0	40	15	0.828
104	0.817	0.000	1.000	0.183	178	0	40	15	0.828
105	0.817	0.000	1.000	0.183	178	0	40	15	0.828
106	0.817	0.000	1.000	0.183	178	0	40	15	0.828

Table A.2 (continued)

cutoff	tprate	fprate	tnrate	fnrate	tp	fp	fn	tn	propcorr
107	0.817	0.000	1.000	0.183	178	0	40	15	0.828
108	0.812	0.000	1.000	0.188	177	0	41	15	0.824
109	0.812	0.000	1.000	0.188	177	0	41	15	0.824
110	0.812	0.000	1.000	0.188	177	0	41	15	0.824
111	0.803	0.000	1.000	0.197	175	0	43	15	0.815
112	0.798	0.000	1.000	0.202	174	0	44	15	0.811
113	0.794	0.000	1.000	0.206	173	0	45	15	0.807
114	0.789	0.000	1.000	0.211	172	0	46	15	0.803
115	0.789	0.000	1.000	0.211	172	0	46	15	0.803
116	0.789	0.000	1.000	0.211	172	0	46	15	0.803
117	0.784	0.000	1.000	0.216	171	0	47	15	0.798
118	0.780	0.000	1.000	0.220	170	0	48	15	0.794
119	0.775	0.000	1.000	0.225	169	0	49	15	0.790
120	0.775	0.000	1.000	0.225	169	0	49	15	0.790
121	0.775	0.000	1.000	0.225	169	0	49	15	0.790
122	0.771	0.000	1.000	0.229	168	0	50	15	0.785
123	0.771	0.000	1.000	0.229	168	0	50	15	0.785
124	0.771	0.000	1.000	0.229	168	0	50	15	0.785
125	0.761	0.000	1.000	0.239	166	0	52	15	0.777
126	0.757	0.000	1.000	0.243	165	0	53	15	0.773
127	0.757	0.000	1.000	0.243	165	0	53	15	0.773
128	0.757	0.000	1.000	0.243	165	0	53	15	0.773
129	0.757	0.000	1.000	0.243	165	0	53	15	0.773
130	0.757	0.000	1.000	0.243	165	0	53	15	0.773
131	0.757	0.000	1.000	0.243	165	0	53	15	0.773
132	0.757	0.000	1.000	0.243	165	0	53	15	0.773
133	0.752	0.000	1.000	0.248	164	0	54	15	0.768

Table A.2 (continued)

cutoff	tprate	fprate	tnrate	fnrate	tp	fp	fn	tn	propcorr
134	0.752	0.000	1.000	0.248	164	0	54	15	0.768
135	0.752	0.000	1.000	0.248	164	0	54	15	0.768
136	0.752	0.000	1.000	0.248	164	0	54	15	0.768
137	0.752	0.000	1.000	0.248	164	0	54	15	0.768
138	0.752	0.000	1.000	0.248	164	0	54	15	0.768
139	0.743	0.000	1.000	0.257	162	0	56	15	0.760
140	0.743	0.000	1.000	0.257	162	0	56	15	0.760
141	0.743	0.000	1.000	0.257	162	0	56	15	0.760
142	0.743	0.000	1.000	0.257	162	0	56	15	0.760
143	0.734	0.000	1.000	0.266	160	0	58	15	0.751
144	0.734	0.000	1.000	0.266	160	0	58	15	0.751
145	0.734	0.000	1.000	0.266	160	0	58	15	0.751
146	0.734	0.000	1.000	0.266	160	0	58	15	0.751
147	0.729	0.000	1.000	0.271	159	0	59	15	0.747
148	0.725	0.000	1.000	0.275	158	0	60	15	0.742
149	0.725	0.000	1.000	0.275	158	0	60	15	0.742
150	0.725	0.000	1.000	0.275	158	0	60	15	0.742
151	0.725	0.000	1.000	0.275	158	0	60	15	0.742
152	0.725	0.000	1.000	0.275	158	0	60	15	0.742
153	0.720	0.000	1.000	0.280	157	0	61	15	0.738
154	0.716	0.000	1.000	0.284	156	0	62	15	0.734
155	0.711	0.000	1.000	0.289	155	0	63	15	0.730
156	0.711	0.000	1.000	0.289	155	0	63	15	0.730
157	0.711	0.000	1.000	0.289	155	0	63	15	0.730
158	0.711	0.000	1.000	0.289	155	0	63	15	0.730
159	0.706	0.000	1.000	0.294	154	0	64	15	0.725
160	0.702	0.000	1.000	0.298	153	0	65	15	0.721

Table A.2 (continued)

cutoff	tprate	fprate	tnrate	fnrate	tp	fp	fn	tn	propcorr
161	0.702	0.000	1.000	0.298	153	0	65	15	0.721
162	0.693	0.000	1.000	0.307	151	0	67	15	0.712
163	0.693	0.000	1.000	0.307	151	0	67	15	0.712
164	0.693	0.000	1.000	0.307	151	0	67	15	0.712
165	0.688	0.000	1.000	0.312	150	0	68	15	0.708
166	0.688	0.000	1.000	0.312	150	0	68	15	0.708
167	0.683	0.000	1.000	0.317	149	0	69	15	0.704
168	0.679	0.000	1.000	0.321	148	0	70	15	0.700
169	0.670	0.000	1.000	0.330	146	0	72	15	0.691
170	0.670	0.000	1.000	0.330	146	0	72	15	0.691
171	0.665	0.000	1.000	0.335	145	0	73	15	0.687
172	0.656	0.000	1.000	0.344	143	0	75	15	0.678
173	0.656	0.000	1.000	0.344	143	0	75	15	0.678
174	0.656	0.000	1.000	0.344	143	0	75	15	0.678
175	0.651	0.000	1.000	0.349	142	0	76	15	0.674
176	0.642	0.000	1.000	0.358	140	0	78	15	0.665
177	0.642	0.000	1.000	0.358	140	0	78	15	0.665
178	0.638	0.000	1.000	0.362	139	0	79	15	0.661
179	0.638	0.000	1.000	0.362	139	0	79	15	0.661
180	0.633	0.000	1.000	0.367	138	0	80	15	0.657
181	0.633	0.000	1.000	0.367	138	0	80	15	0.657
182	0.633	0.000	1.000	0.367	138	0	80	15	0.657
183	0.633	0.000	1.000	0.367	138	0	80	15	0.657
184	0.633	0.000	1.000	0.367	138	0	80	15	0.657
185	0.633	0.000	1.000	0.367	138	0	80	15	0.657
186	0.633	0.000	1.000	0.367	138	0	80	15	0.657
187	0.633	0.000	1.000	0.367	138	0	80	15	0.657

Table A.2 (continued)

cutoff	tprate	fprate	tnrate	fnrate	tp	fp	fn	tn	propcorr
188	0.633	0.000	1.000	0.367	138	0	80	15	0.657
189	0.633	0.000	1.000	0.367	138	0	80	15	0.657
190	0.633	0.000	1.000	0.367	138	0	80	15	0.657
191	0.633	0.000	1.000	0.367	138	0	80	15	0.657
192	0.628	0.000	1.000	0.372	137	0	81	15	0.652
193	0.628	0.000	1.000	0.372	137	0	81	15	0.652
194	0.628	0.000	1.000	0.372	137	0	81	15	0.652
195	0.628	0.000	1.000	0.372	137	0	81	15	0.652
196	0.628	0.000	1.000	0.372	137	0	81	15	0.652
197	0.628	0.000	1.000	0.372	137	0	81	15	0.652
198	0.619	0.000	1.000	0.381	135	0	83	15	0.644
199	0.619	0.000	1.000	0.381	135	0	83	15	0.644
200	0.615	0.000	1.000	0.385	134	0	84	15	0.639
201	0.615	0.000	1.000	0.385	134	0	84	15	0.639
202	0.615	0.000	1.000	0.385	134	0	84	15	0.639
203	0.615	0.000	1.000	0.385	134	0	84	15	0.639
204	0.615	0.000	1.000	0.385	134	0	84	15	0.639
205	0.610	0.000	1.000	0.390	133	0	85	15	0.635
206	0.610	0.000	1.000	0.390	133	0	85	15	0.635
207	0.610	0.000	1.000	0.390	133	0	85	15	0.635
208	0.606	0.000	1.000	0.394	132	0	86	15	0.631
209	0.606	0.000	1.000	0.394	132	0	86	15	0.631
210	0.606	0.000	1.000	0.394	132	0	86	15	0.631
211	0.606	0.000	1.000	0.394	132	0	86	15	0.631
212	0.606	0.000	1.000	0.394	132	0	86	15	0.631
213	0.606	0.000	1.000	0.394	132	0	86	15	0.631
214	0.606	0.000	1.000	0.394	132	0	86	15	0.631

Table A.2 (continued)

cutoff	tprate	fprate	tnrate	fnrate	tp	fp	fn	tn	propcorr
215	0.606	0.000	1.000	0.394	132	0	86	15	0.631
216	0.606	0.000	1.000	0.394	132	0	86	15	0.631
217	0.606	0.000	1.000	0.394	132	0	86	15	0.631
218	0.606	0.000	1.000	0.394	132	0	86	15	0.631
219	0.606	0.000	1.000	0.394	132	0	86	15	0.631
220	0.606	0.000	1.000	0.394	132	0	86	15	0.631
221	0.606	0.000	1.000	0.394	132	0	86	15	0.631
222	0.606	0.000	1.000	0.394	132	0	86	15	0.631
223	0.606	0.000	1.000	0.394	132	0	86	15	0.631
224	0.606	0.000	1.000	0.394	132	0	86	15	0.631
225	0.601	0.000	1.000	0.399	131	0	87	15	0.627
226	0.601	0.000	1.000	0.399	131	0	87	15	0.627
227	0.601	0.000	1.000	0.399	131	0	87	15	0.627
228	0.601	0.000	1.000	0.399	131	0	87	15	0.627
229	0.601	0.000	1.000	0.399	131	0	87	15	0.627
230	0.596	0.000	1.000	0.404	130	0	88	15	0.622
231	0.596	0.000	1.000	0.404	130	0	88	15	0.622
232	0.596	0.000	1.000	0.404	130	0	88	15	0.622
233	0.596	0.000	1.000	0.404	130	0	88	15	0.622
234	0.596	0.000	1.000	0.404	130	0	88	15	0.622
235	0.596	0.000	1.000	0.404	130	0	88	15	0.622
236	0.592	0.000	1.000	0.408	129	0	89	15	0.618
237	0.592	0.000	1.000	0.408	129	0	89	15	0.618
238	0.592	0.000	1.000	0.408	129	0	89	15	0.618
239	0.587	0.000	1.000	0.413	128	0	90	15	0.614
240	0.587	0.000	1.000	0.413	128	0	90	15	0.614
241	0.583	0.000	1.000	0.417	127	0	91	15	0.609

Table A.2 (continued)

cutoff	tprate	fprate	tnrate	fnrate	tp	fp	fn	tn	propcorr
242	0.583	0.000	1.000	0.417	127	0	91	15	0.609
243	0.583	0.000	1.000	0.417	127	0	91	15	0.609
244	0.573	0.000	1.000	0.427	125	0	93	15	0.601
245	0.573	0.000	1.000	0.427	125	0	93	15	0.601
246	0.564	0.000	1.000	0.436	123	0	95	15	0.592
247	0.564	0.000	1.000	0.436	123	0	95	15	0.592
248	0.564	0.000	1.000	0.436	123	0	95	15	0.592
249	0.564	0.000	1.000	0.436	123	0	95	15	0.592
250	0.564	0.000	1.000	0.436	123	0	95	15	0.592
251	0.564	0.000	1.000	0.436	123	0	95	15	0.592
252	0.564	0.000	1.000	0.436	123	0	95	15	0.592
253	0.564	0.000	1.000	0.436	123	0	95	15	0.592
254	0.564	0.000	1.000	0.436	123	0	95	15	0.592
255	0.564	0.000	1.000	0.436	123	0	95	15	0.592
256	0.564	0.000	1.000	0.436	123	0	95	15	0.592
257	0.564	0.000	1.000	0.436	123	0	95	15	0.592
258	0.564	0.000	1.000	0.436	123	0	95	15	0.592
259	0.564	0.000	1.000	0.436	123	0	95	15	0.592
260	0.560	0.000	1.000	0.440	122	0	96	15	0.588
261	0.560	0.000	1.000	0.440	122	0	96	15	0.588
262	0.555	0.000	1.000	0.445	121	0	97	15	0.584
263	0.550	0.000	1.000	0.450	120	0	98	15	0.579
264	0.546	0.000	1.000	0.454	119	0	99	15	0.575
265	0.546	0.000	1.000	0.454	119	0	99	15	0.575
266	0.546	0.000	1.000	0.454	119	0	99	15	0.575
267	0.541	0.000	1.000	0.459	118	0	100	15	0.571
268	0.537	0.000	1.000	0.463	117	0	101	15	0.567

Table A.2 (continued)

cutoff	tprate	fprate	tnrate	fnrate	tp	fp	fn	tn	propcorr
269	0.537	0.000	1.000	0.463	117	0	101	15	0.567
270	0.537	0.000	1.000	0.463	117	0	101	15	0.567
271	0.537	0.000	1.000	0.463	117	0	101	15	0.567
272	0.537	0.000	1.000	0.463	117	0	101	15	0.567
273	0.537	0.000	1.000	0.463	117	0	101	15	0.567
274	0.537	0.000	1.000	0.463	117	0	101	15	0.567
275	0.537	0.000	1.000	0.463	117	0	101	15	0.567
276	0.537	0.000	1.000	0.463	117	0	101	15	0.567
277	0.537	0.000	1.000	0.463	117	0	101	15	0.567
278	0.537	0.000	1.000	0.463	117	0	101	15	0.567
279	0.537	0.000	1.000	0.463	117	0	101	15	0.567
280	0.537	0.000	1.000	0.463	117	0	101	15	0.567
281	0.537	0.000	1.000	0.463	117	0	101	15	0.567
282	0.528	0.000	1.000	0.472	115	0	103	15	0.558
283	0.528	0.000	1.000	0.472	115	0	103	15	0.558
284	0.528	0.000	1.000	0.472	115	0	103	15	0.558
285	0.523	0.000	1.000	0.477	114	0	104	15	0.554
286	0.518	0.000	1.000	0.482	113	0	105	15	0.549
287	0.518	0.000	1.000	0.482	113	0	105	15	0.549
288	0.518	0.000	1.000	0.482	113	0	105	15	0.549
289	0.518	0.000	1.000	0.482	113	0	105	15	0.549
290	0.518	0.000	1.000	0.482	113	0	105	15	0.549
291	0.514	0.000	1.000	0.486	112	0	106	15	0.545
292	0.514	0.000	1.000	0.486	112	0	106	15	0.545
293	0.514	0.000	1.000	0.486	112	0	106	15	0.545
294	0.514	0.000	1.000	0.486	112	0	106	15	0.545
295	0.514	0.000	1.000	0.486	112	0	106	15	0.545

Table A.2 (continued)

cutoff	tprate	fprate	tnrate	fnrate	tp	fp	fn	tn	propcorr
296	0.514	0.000	1.000	0.486	112	0	106	15	0.545
297	0.509	0.000	1.000	0.491	111	0	107	15	0.541
298	0.500	0.000	1.000	0.500	109	0	109	15	0.532
299	0.495	0.000	1.000	0.505	108	0	110	15	0.528
300	0.495	0.000	1.000	0.505	108	0	110	15	0.528

REFERENCES

Bowie, W.B., King A.S., Werker, D.H., Isaac-Renton, J.L., Bell, A., Marion, S.A. 1997. Outbreak of toxoplasmosis associated with municipal drinking water. *Lancet* 350:173-177.

Chappell, C.L., Okhuysen, P.C., Sterling, C.R., Wang, C., Jakubowski, W., Dupont, H.L. 1999. Infectivity of Cryptosporidium parvum in healthy adults with pre-existing anti-C. parvum serum immunoglobulin G. Am J Trop Med Hyg 1999 60(1):157-164.

Chen, W., Harp, J.A., Harmsen, A.G., Havell, E.A. 1993. Gamma interferon functions in resistance to *Cryptosporidium parvum* infection in severe combined immunodeficient mice. *Infect Immun* 61(8):3548-3551.

Clancy, J.L. 2000. Sydney's 1998 water quality crisis. *Jour. AWWA* 92(3):55-66.

Craun, G.F., Hubbs, S.A., Frost, F., Calderon, R., Via, S. 1998. Waterborne outbreaks of cryptosporidiosis. *Jour. AWWA* 90(9):81-91.

Current, W.L., Reese, N.C., Ernst, J.V., Bailey, W.S., Heyman, M.B., Weinstein, W.M. 1983. Human cryptosporidiosis in immunocompetent and immunodeficient persons: Studies of an outbreak and experimental transmission. *N Engl J Med* 308(21):1252-1257.

Doyle, P.S., Crabb, J., Petersen, C. 1993. Anti-*Cryptosporidium parvum* antibodies inhibit infectivity in vitro and in vivo. *Infect Immun* 61(10):4079-4084.

Dunnett, C.W. 1955. A multiple comparison procedure for comparing several treatments with a control. *J Amer Stat Assoc* 50:1096-1121.

Dunnett, C.W. 1964. New tables for multiple comparison with a control. *Biometrics* 20: 482-491

DuPont, H.L., Chappell, C.L., Sterling, C.R., Okhuysen, P.C., Rose, J.B., Jakubowski, W. 1995. The infectivity of *Cryptosporidium parvum* in healthy volunteers. *N Eng J Med* 332(13):855-859.

Fayer, R., Perryman, L.E., Riggs, M.W. 1989. Hyperimmune bovine colostrum neutralizes *Cryptosporidium* sporozoites and protects mice against oocyst challenge. *J Parasitol* 75(1):151-153.

Fayer, R., Guidry, A., Blagburn, B.L. 1990. Immunotherapeutic efficacy of bovine colostral immunoglobulins from a hyperimmunized cow against cryptosporidiosis in neonatal mice. *Infect Immun* 58(9):2962-2965.

Flanigan, T., Whalen, C., Turner, J., Soave, R., Toerner, J., Havlir, D., Kotler, D. 1992. *Cryptosporidium* infection and CD4 counts. *Ann Intern Med* 116(10):840-842.

Frost, F.J., de la Cruz, A.A., Moss, D.M., Curry, M., Calderon, R.L. 1998a. Comparison of ELISA and Western blot assays for detection of *Cryptosporidium* antibody. *Epidemiol Infect* 121(1):205-211.

Frost, F.J., Calderon, R.L., Muller, T.B., Curry, M., Rodman, J.S. Moss, D.M., de la Cruz, A.A. 1998b. A two-year follow-up survey of antibody to *Cryptosporidium* in Jackson County, Oregon, following an outbreak of waterborne disease. *Epidemiol Infect* 121(1):213-217.

Garcia, L.S. and Bruckner, D.A. 1997. *Diagnostic Medical Parasitology.* (3rd Ed.) American Society of Microbiology, Washington D.C.

Gregory, M.W. 1990. Epidemiology of cryptosporidiosis in animals. Cryptosporidiosis in water supplies. Report of the group of experts (chairman Sir John Badenoch). HM Stationery Office, London.

Griffiths, J.K., Theodos, C., Paris, M., Tzipori, S. 1998. The gamma interferon gene knockout mouse: a highly sensitive model for evaluation of therapeutic agents against *Cryptosporidium parvum*. *J Clin Microbiol* 36(9):2503-2508.

Hayes, E.B., Matte, T.D., O'Brien, T.R., McKinley, T.W., Logsdon, G.S., Rose, J.B., Ungar, B.L.P., Word, D.M., Pinsky, P.F., Cummings, M.L., Wilson, M.A., Long, E.G., Hurwitz, E.S., Juranek, D.D. 1989. Large community outbreak of cryptosporidiosis due to contamination of a filtered public water supply. *N Eng J Med* 320(21):1372-1376.

Isaac-Renton, J.L., Lewis, L.F., Ong, C.S.L., Nulsen, M.F. 1994. A second community outbreak of waterborne giardiasis in Canada and serological investigation of patients. *Trans R Soc Trop Med Hyg* 88(4):395-399.

Isaac-Renton, J.L., Blatherwick, J., Bowie, W.R., Fyfe, M., Khan, M., Li, a., King, A., McLean, M., Medd, L., Moorehead, W., Ong, C., Robertson, W. 1999. Epidemic and endemic seroprevalence of antibodies to *Cryptosporidium* and *Giardia* in residents of three communities with different drinking water supplies. *Am J Trop Med Hyg* 60(4):578-583.

LeChevallier, M.W., Norton, W.D., Lee, R.G. 1991. Occurrence of *Giardia* and *Cryptosporidium spp*. in surface water supplies. *Appl Environ Microbiol* 57(9):2610-2616.

Leland, D., McAnulty, J., Keene, W., Stevens, G. 1993. A cryptosporidiosis outbreak in filtered-water supply. *Jour. AWWA* 85:34-42.

MacKenzie, W., Neil, M., Hoxie, N., Proctor, M., Gradus, M., Blair, K., Peterson, D., Kazmierczak, J., Addiss, D., Fox, K., Rose, J., Davis, J. 1994. A massive outbreak in Milwaukee of *Cryptosporidium* infection transmitted through the public water supply. *N Eng J Med* 331 (3) : 161-167.

Mead, J.R., Arrowood, M.J., Sidwell, R.W., Healey, M.C. 1991. Chronic *Cryptosporidium parvum* infections in congenitally immunodeficient SCID and nude mice. *J Infect Dis* 163(6): 1297-1304.

Metz, C.E. 1978. Basic principles of ROC analysis. *Semin Nucl Med* 8(4):283-298.

Moss, D.M., Bennett, S.N., Arrowood, M.J., Hurd, M.R., Lammie, P.J., Aahlquist, S.P., Addiss, D.G. 1994. Kinetic and isotypic analysis of specific immunoglobulins from crew members with cryptosporidiosis on a U.S. coast guard cutter. *J Eukaryot Microbiol* 41(5):52S-55S.

Moss, D.M., Chappell, C.L., Okhuysen, P.C., DuPont, H.L., Arrowood, M.J., Hightower, A.W., Lammie, P.J. 1998. The antibody response to 27-, 17-, and 15-kDa *Cryptosporidium* antigens following experimental infection in humans. *J Infect Dis* 178 (3): 827-833.

Nord, J., Ma, P., DiJohn, D., Tzipori, S., Tacket, C.O. 1990. Treatment with bovine hyperimmune colostrum of cryptosporidial diarrhea in AIDS patients. *AIDS* 4(6):581-584.

Petry, F. 1998. Epidemiological study of *Cryptosporidium parvum* antibodies in sera of persons from Germany. *Infection* 26(1):7-10.

Priest, J.W., Kwon, J.P., Moss, D.M., Roberts, J.M., Arrowood. M.J., Dworkin, M.S., Juranek, D.D., Lammie, P.J.1999. Detection by enzyme immunoassay of serum immunoglobulin G antibodies that recognize specific *Cryptosporidium parvum* antigens. *J Clin Microbiol* 37(5): 1385-1392.

Priest, J.W., Li, A., Khan, M., Arrowood, M.J., Lammie, P.J., Ong, C.S., Roberts, J.M., Isaac-Renton, J. 2001. Enzyme immunoassay detection of antigen-specific immunoglobulin G antibodies in longitudinal serum samples from patients with cryptosporidiosis. Clin Diagn Lab Immunol 8(2):415-423

Richardson, A.J., Frankenberg, R.A., Buck, A.C., Selkon, J.B., Colbourne, J.S., Parsons, J.W., Mayon-White, R.T. 1991. An outbreak of waterborne cryptosporidiosis in Swindon and Oxfordshire. *Epidemiol Infect* 107(3):485-495.

Shewhart, W.A. 1931. *Economic control of quality of manufactured product.* New York: Van Nostrand.

Smith, L.M., Bonafonte, M.T., Mead, J.R. 2000. Cytokine expression and specific lymphocyte proliferation in two strains of *Cryptosporidium parvum*-infected gamma-interferon knockout mice. *J Parasitol* 86(2):300-307.

Taghi-Kilani, R., Sekla, L., Hayglass, K.T. 1990. The role of humoral immunity in *Cryptosporidium* spp. infection. Studies with B-cell depleted mice. *J Immunol* 145(5):1571-1576.

Tzipori, S., Roberton, D., Chapman, C. 1986. Remission of diarrhoea due to cryptosporidiosis in an immunodeficient child treated with hyperimmune bovine colostrum. *Br Med J* 293(6557): 1276-1277.

Tzipori, S. 1988. Cryptosporidiosis in perspective. *Adv Parasitol* 27:63-129.

Ungar, B.L.P., Soave, R, Fayer, R, Nash, T.E. 1986. Enzyme immunoassay detection of immunoglobulin M and G antibodies to *Cryptosporidium* in immunocompetent and immunocompromised persons. *J Infect Dis* 153(3):570-578.

Ungar, B.L.P., Mulligan, M., Nutman, T.B. 1989. Serologic evidence of *Cryptosporidium* infection in US volunteers before and during Peace Corps service in Africa. *Arch Intern Med* 149(4):894-897.

Ungar, B.L.P., Ward, D.J., Fayer, R., Quinn, C.A. 1990a. Cessation of *Cryptosporidium*-associated diarrhea in an acquired immunodeficiency syndrome patient after treatment with hyperimmune bovine colostrum. *Gastroenterology* 98(2):486-489.

Ungar, B.L.P., Burris, J.A., Quinn, C.A., Finkelman, F.D. 1990b. New mouse models for chronic *Cryptosporidium* infection in immunodeficient hosts. *Infect Immun* 58(4):961-969.

Ungar, B.L.P., Kao, T-C., Burris, J.A., Finkelman, F.D. 1991. *Cryptosporidium* infection in an adult mouse model. Independent roles for IFN-g and CD4+ T lymphocytes in protective immunity. *J Immunol* 147(3):1014-1022.

Westgard, J.O., Barry, P.I, Hunt, M.R., Groth, T. 1981. A multi-rule Shewhart chart for quality control in clinical chemistry. *Clin Chem* 27(3):493-501.

Zu, S.X., Fang, G.D., Fayer, R., Guerrant, R.L. 1992. Cryptosporidiosis: pathogenesis and immunology. *Parasitology Today* 8:24-27.

Zu, S.X., Li, J.F., Barrett, L.J., Fayer, R., Shu, S.Y., McAuliffe, J.F., Roche, J.K., Guerrant, R.L. 1994. Seroepidemiologic study of *Cryptosporidium* infection in children from rural communities of Anhui, China and Fortaleza, Brazil. *Am J Trop Med Hyg* 51(1):1-10.

Zweig, M.H., Campbell, G. 1993. Receiver-operating characteristic (ROC) plots: a fundamental evaluation tool in clinical medicine (review). *Clin Chem* 39(4):561-577.

ABBREVIATIONS

AWWA	American Water Works Association
AwwaRF	Awwa Research Foundation
ASTM	American Society for Testing and Materials
Avg	average
BC	British Columbia
BCA	bicinchoninic acid
BCCDC	British Columbia Centre for Disease Control
°C	degrees Celcius (centigrade)
CDC	Centers for Disease Control
Cp23	recombinant form of the C. parvum 27 kDa antigen
cutoff	threshold for positivity in arbitrary units
CV	coefficient of variation
EITB	enzyme-linked immunoelectrotransfer blot
ELISA	enzyme-linked immunosorbent assay
fn	number of false negatives
fnrate	false negative rate
fp	number of false positives
fprate	false positive rate
GST	Schistosoma japonicum glutathione-S-transferase
IgA	immunoglobulin A

IgG	immunoglobulin G
IgM	immunoglobulin M
kDa	kilodalton
L	liter
LCL	lower control limit
NC	not calculated
PAC	Project Advisory Committee
pGEX	expression vector from Pharmacia Biotech
propcorr	proportion correct
SAF	sodium acetate-acetic acid-formalin
SCID	severe combined immune deficiency
SD	standard deviation
SDS	sodium dodecyl sulphate
TIFF	tagged image format file
tn	number of true negatives detected at the given cutoff
tnrate	true negative rate
tp	number of true positives detected at the given cutoff value
tprate	true positive rate
U	units
UBC	University of British Columbia
UCL	upper control limit

IWA Publishing's authorised EU representative for General Product Safety Regulations is Diane D'Arras, 15 rue Duret, 75116 Paris, France, e-mail: safety@iwap.co.uk.

Printed and bound by CPI Group (UK) Ltd, Croydon, CR0 4YY
17/09/2026
02218808-0011